VIDEO EDITING

AND POST-PRODUCTION:
A PROFESSIONAL GUIDE

2nd Edition

GARY H. ANDERSON

Knowledge Industry Publications, Inc.
White Plains, NY

Video Bookshelf

Video Editing and Post-Production:
A Professional Guide, Second Edition

Library of Congress Cataloging in Publication Data

Anderson, Gary H.
    Video editing and post-production.

    (Video bookshelf)
    Bibliography: p.
    Includes index.
    1. Video tapes—Editing. I. Title. II. Series.
TR899.A57 1988   778.59'92   87-33925
ISBN 0-86729-257-1
ISBN 0-86729-258-X (pbk.)

Printed in the United States of America

10 9 8 7 6 5 4 3

Contents

Figures and Tables

Acknowledgments

I would like to express my sincere thanks to Loren Lundt and Clyde Spear for their general advice and invaluable technical expertise and to Burton I. Lippman, Newton Bellis and Rita Scott for their generous support.

For the use of photographs and data, thanks and appreciation go to: Alpha Audio; Ampex Corp.; CMX Corp.; Calaway Engineering; Capitol Records, Inc.; Comprehensive Video Supply Corp.; Qualogy, Inc.; EECO, Inc.; For.A Corp.; Fostex Corporation of America; The Grass Valley Group, Inc.; JVC Co. of America; Laseredit, Inc.; Lexicon, Inc.; New England Digital; Omnimusic; Panasonic Industrial Co.; The Post Group; Sony Communications Products Co.; TAV Sound; Tektronix, Inc.; Unitel Video; Videomedia, Inc.; Vidtronics, Inc.; and Vital Industries, Inc.

One Day at a Time photographs are used by permission © copyright Embassy Television, all rights reserved. My thanks also to Patricia Fass Palmer and to Valerie Bertinelli, Bonnie Franklin and Pat Harrington.

I am indebted to Phil Miller for his fine efforts and expertise in transforming my manuscript into this book and to the following individuals for their comments and suggestions: Denny Allen, Dave Bargen, Dr. Richard Bluthian, Jerry Clemans, Jim Conlon, Doug Davies, Richard B. Davies, Karen Drew, Jim Van Eaton, Richard Garibaldi, Bill Gordon, Tamara Johnson, Sedhae Kim and Stanley Perkins.

Finally, I wish to express my deepest gratitude to my wife, Diane and to my sons, Darrell and Craig, who encouraged me throughout the preparation of both of these editions. My wife Diane wishes to express her deepest gratitude to Apple Computer, Inc. because she didn't have to hand type the new edition.

Introduction

A revolution continues in the television industry. Since the mid-1970s cable television, home video recorders, subscription TV services, interactive laser discs and corporate training videos have been creating new outlets for entertainment, educational and information programming, and new opportunities for program producers and distributors. At the same time, technical advances in video recording, signal processing and electronic editing equipment have transformed the TV production process, creating new options for both film and video directors, editors and production specialists.

The aspects of the video industry that have probably undergone the most radical changes are the video editing and audio post-production stages. In less than three decades, video editing has evolved from a crude "cut and splice" exercise into a highly sophisticated, computerized craft. In the same time span, audio post-production for video has begun to rival film sound techniques in its quality and complexity. In recent years, video editing and audio post-production have also emerged as critical components in the TV production process—the final steps that shape the look, pace and "feel" of the finished television program.

Curiously enough, however, video and audio post-production still remain the least understood areas of the video industry. To many directors and producers, post-production is still a mysterious process that takes place in the dark confines of the editing bay or mixing suite, where shadowy figures perform magical tricks that transform raw footage into a polished TV program. Often, there is little significant communication

and cooperation between production and post-production professionals. This situation makes it difficult for producers and directors to anticipate and avoid the many problems that can appear once a project enters the editing stage.

I wrote the first edition of *Video Editing and Post-Production: A Professional Guide* with this special set of problems in mind. The tremendously gratifying national and international response to that edition, plus the comments from many students and professors, indicates to me that progress is being made. In video post-production, as in any production process, there is a logical sequence of steps to follow to create the best possible product. By identifying and explaining these steps, I hope to continue to help dissipate the cloud of confusion and misunderstanding that surrounds video post-production.

Although it does include some technical information, this second edition of *Video Editing and Post-Production: A Professional Guide* is not intended as a technical or engineering guide, nor as an operator's manual for specific editing systems. In writing the technical sections, I have been careful to emphasize the basic information that professionals need to acquire a practical, working knowledge of editing equipment and the post-production process.

This book begins with a brief history of video editing technology. Chapter 1 also provides an introduction to the art and technique of video editing—selecting shots, constructing continuity, and determining pace and timing. Chapter 2, the most technical chapter, covers the basic technical information and video recording formats editors need to be aware of in order to understand video post-production. In Chapter 3, readers take a tour of a typical editing bay, with stops to introduce the various components used in preparing, performing and monitoring videotape edits.

Chapters 4 through 6 cover the actual post-production process: preparation for editing, offline editing and online editing. Chapter 4 includes information on budgeting, videotape stock selection, transference of film to videotape and delegation of responsibility among members of the post-production team. In Chapter 5, readers receive an overview of offline editing—the critical stage in which editors prepare video workprints and an edit decision list. In Chapter 6, readers learn how to use the edit decision list to create a finished video project in a computerized, online editing session.

Chapter 7 provides an overview of digital special effects, a fast-growing area of post-production that provides editors and producers with exciting new options for creating video shape manipulations, moving image effects and scene transitions.

In this second edition, I have included an additional chapter which discusses the second most confusing aspect of video post-production—audio post-production. Chapter 8 covers an overall view of audio post-production for video. In this chapter, readers will take a tour through an audio facility to become familiar with the various pieces of equipment and what they can do. Processes such as spotting music and sound effects, automatic dialogue replacement (ADR), foley sound effects, multitrack music mixdowns and digital recording techniques will help demonstrate the great potential that audio post-production has to offer the film or video producer. Finally, Chapter 9 looks to the future of video post-production techniques and technologies.

In writing each chapter, I have drawn heavily on my years of practical experience in television production and video editing. I became a video editor in the late 1960s, before computerized editing systems were introduced to the television industry. As a result, I have had the opportunity to grow with computerized videotape editing and to witness the peaks and valleys of a maturing technology.

Using this experience, I have tried to present information that will benefit both novices and established editing professionals. For the novice, I have included basic information on editing equipment and techniques. For the established professional, I have included tips on budgeting, organization of post-production sessions and how to get the best possible performance from the equipment and personnel. It is my hope, that this second edition of *Video Editing and Post-Production: A Professional Guide* will continue to serve both as a thorough introductory text for readers just getting into video editing and as a valuable handbook and reference for experienced editing professionals.

1 The History, Technology and Technique of Video Editing

Post-production encompasses all of the activities necessary to transform raw video footage into a finished TV program. In most professional TV productions, those activities include the following: logging the raw footage, determining edit points and transitions, generating edit decision lists, assembling the final master, adding titles and graphics, and, finally, adding sound effects and music. Technically, editing is only one step in the post-production process—the step in which selected shots and segments are electronically pieced together to form a rough workprint or a finished TV program.

Some television programs require very little post-production processing. This is sometimes the case with TV game shows or talk shows that are videotaped "live," using multiple cameras and a video switcher, and that are edited only for program length (see Figure 1.1). In general, however, post-production is as necessary in producing video programs as it is in producing films and for many of the same reasons. Actors' performances may need to be edited for consistency, pacing may need to be improved, camera angles may need to be changed, and music and sound effects may need to be added. In fact, in dramatic or documentary programs shot at several locations, the entire program is essentially put together in post-production.

In subsequent chapters, I have outlined the various steps involved in carrying a program through post-production. For the most part, video post-production mirrors the editing and post-production process used in film, except that video producers are generally able to avoid the extra time and expense required to develop film workprints. In this chapter, you will learn how video editing was born and how technological developments and the evolving needs of TV producers have brought video post-production to its current computerized state. You will also learn about the art of video editing—the artistic principles that guide a video editor in the selection of shots, camera angles and transitions.

A SHORT HISTORY OF VIDEO EDITING TECHNOLOGY

The history of video editing is closely tied to the history of videotape recording. By most accounts, video recording was born in 1956, when Ampex Corp. unveiled the first 2-inch broadcast videotape recorder (VTR) at the annual convention of the National Association of Radio and Television Broadcasters. To design the machine, Ampex had organized a team of engineers that included Charles Anderson, the "father of FM radio," and Ray Dolby, later to gain fame for the Dolby noise reduction system (see Figure 1.2).

The response of the TV industry was overwhelming. In November 1956, the first videotape recorder went "on air" from Television City in Hollywood, and Ampex was soon swamped with orders for the new machines. Most of the orders came from the broadcast networks and their affiliated TV stations, who needed the VTRs for recording and rebroadcasting "network delay programming." Since most network broadcasts

1

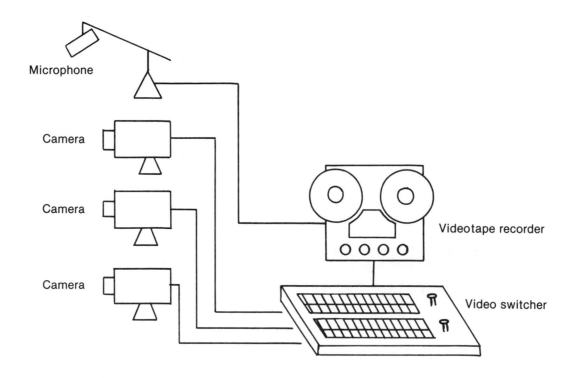

Figure 1.1:   A simple method of taping a program "live," using multiple cameras and a video switcher.

originated in New York, the new video recorders were especially popular with West Coast affiliates, who needed some way of recording and replaying programming to compensate for the three-hour time difference. Before the advent of VTRs, most West Coast stations would film the programs from a TV monitor, using an expensive and technically inferior process called "kinescoping."

Although this "net delay" application proved to be the primary use of early VTRs, producers and video engineers soon sensed that VTRs could be used in TV production, too. At the time, TV programs were either broadcast live, using the "switch-feed" method shown in Figure 1.1, or produced and edited on film. For TV producers, VTRs promised to provide an inexpensive alternative to film and a convenient means of recording and "tightening up" switchfed productions—as soon as someone figured out how to edit the videotape.

Slice and Splice

In the late 1950s, the only practical method of editing videotape involved physically slicing and splicing

the tape. To facilitate this process, Ampex developed and marketed a videotape splicing block in 1958. Several other companies soon followed its lead, and professional videotape editing was born.

One of the more common splicing blocks is shown in Figure 1.3. It consisted of a 2-inch wide tape guide (to accommodate the 2-inch wide quadruplex videotape), a precision cutting blade mounted across the guide and a small microscope for viewing the splice point. To operate the splicer, the editor would first apply an iron particle solution (called EDIVUE) over the designated splice area. This would "develop" the video tracks, which would show up under the microscope as a series of vertical strips running across the length of the tape (see Figure 1.4). In addition, the iron particle solution would develop the "frame pulses" that coincided with the start of each video frame.

To perform an edit, the editor would determine the proper edit point, line up the appropriate frame pulse mark under the microscope and then use the precision blade to cut the videotape between the second and third video stripes to the right of the pulse mark. Finally, the editor would carefully splice two sections

Figure 1.2: The six-man team that developed the first practical videotape recorder displays the Emmy Award won by Ampex Corp. in 1957. From left to right, Charles E. Anderson, Ray Dolby, Alex Maxey, Shelby Henderson, Charles Ginsburg and Fred Pfost. Courtesy of Ampex Corp.

of videotape, using an ultra-thin metallic adhesive tape and making sure there was no gap between the two pieces of videotape at the splice point.

Although this mechanical method seems crude by today's standards, it was used very successfully for years on numerous network shows, including "The Dinah Shore Chevy Show" and the original "Rowan and Martin's Laugh-In."

As TV productions became more elaborate, so did the methods of mechanically splicing videotape. For example, to produce a 1959 program that included a medley of songs shot at several locations, two CBS engineers developed a special technique for synchronizing the video segments with a prerecorded audio track. Not to be outdone, NBC developed a "double system" in which a production session was simultaneously recorded on videotape, 16mm kinescope film and 16mm magnetic sound recorders. Paramount Television Productions also developed its own mechanical editing system, as a way of appeasing film producers who were uncomfortable about not

being able to see the video frames as they were determining edit points.

Despite the many advantages offered by later electronic and computerized editing systems, various forms of physical splicing are still being used today, especially in network situations where a program must be quickly edited for time or content and then aired three hours later.

Electronic Editing

In the early 1960s, the first electronic editing device appeared on the market. It allowed shots and scenes to be pieced together without physically cutting the videotape. This new editing technique required editors to play back the tape or tapes containing original shots and scenes on one VTR (commonly called the "playback" or "source" VTR), and to rerecord the shots and scenes in the desired sequence on a second VTR (commonly called the "record" or "edit" VTR). With one of the new electronic editing devices installed

Figure 1.3: A common mechanical videotape splicer known as the Smith splicer. Photo by Darrell R. Anderson.

in the record VTR, the shots and scenes from the source tapes could be pieced together without distortion or ''break-up'' at the edit points. In other words, even in its earliest form, electronic editing required editors to select, sequence and rerecord video material, much as we do today.

Electronic editing presented two primary advantages over mechanical splicing. First, since the edits were performed electronically, damage due to physical handling of the tape was reduced. Second, since electronic editing eliminated the need to cut the tape, the original master tape was preserved. This allowed editors to use scenes repeatedly, if desired, and to repeat and correct unsatisfactory edits.

The one serious disadvantage stemmed from the technical limitations of the early editing devices. Since the systems were operated manually (automatic editing controllers came later), the quality and accuracy of the edit depended solely on the accuracy of the person punching the record button when the tapes from the

source and record VTRs reached the edit points. As a result, electronic editing, in its early stage, was fondly referred to by many as the ''punch and crunch'' method.

Needless to say, these early systems did not allow editors to preview edits. With no way of previewing edits and with the limited accuracy of the punch and crunch method, working with one of the early electronic editing systems could be a frustrating experience. What was needed was a way of presetting the edit points on the tape, so the edit could be previewed, and changes could be made before the edit was actually performed.

Keep in mind that electronic editing is a dynamic process. For each edit, videotapes on at least two VTRs must be rolling, synchronized and up to the correct recording speed (''up-to-speed'') when they reach the edit point (the point at which the circuitry in the record VTR is activated to record the signal sent from the source VTR). In order to synchronize

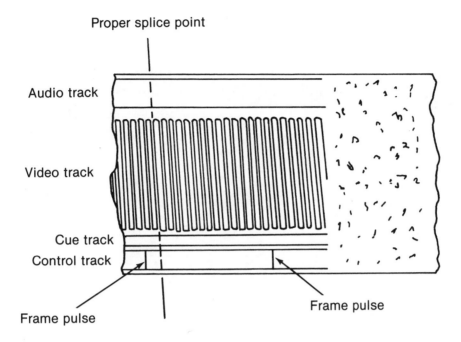

Proper splice point

Audio track

Video track

Cue track
Control track

Frame pulse

Frame pulse

Figure 1.4: Partial section of a developed 2-inch videotape recording showing the proper spot for physically editing the tape.

the tapes on both the source and the record VTRs, many editors began to use white grease pencils to mark the back of the videotape as a way of indicating the correct cue-up points. Most editors placed the cue marks at a spot exactly ten seconds before the edit points, so both tapes were synchronized and fully up to speed when they reached the edit points. The ''grease pencil method'' also allowed editors to preview edits and to change the pencil markings if adjustments were necessary.

In 1963, Ampex Corp. came out with a new editing control device that marked the tape electronically (see Figure 1.5). Called EDITEC, it placed a brief audio tone on the record VTR's cue track as a way of activating the recorder's electronic editing circuitry. Like the earlier Ampex editing device, EDITEC contained controls that allowed audio-only, video-only, or audio and video edits. In addition, EDITEC offered an advance/delay cue system that allowed frame-by-frame shifting of edit points and an ''automatic'' edit mode that made videotape animation possible.

Although EDITEC gave editors greater control, video editing was still an imprecise and somewhat awkward process. As one way of improving the process, many video editors, including the author, began

to use remote VTR start buttons. These devices served two purposes. First, since the bulk and size of the 2-inch VTRs made it difficult to reach the play buttons on both machines simultaneously, the remote control buttons made it much easier for editors to synchronize the start of the source and edit VTRs. Second, with the start controls for the two VTRs nearby, it was much easier for editors to manage the other equipment they were often responsible for operating—equipment that included audio mixers, video production switchers, and film and slide chains.

Time Code Editing

While most editors were still struggling with grease pencils and remote-start buttons, a TV engineer named Dick Hill asked a question that would eventually change the entire video industry: Why not use an electronic time code for video editing, much in the way that edge numbers are used in film editing?

Picking up on Hill's idea, the Electronic Engineering Company of America (now EECO, Inc.) developed the EECO Time Code System. Introduced in 1967, EECO time code employed a technique similar to the method NASA used to encode the telemetry

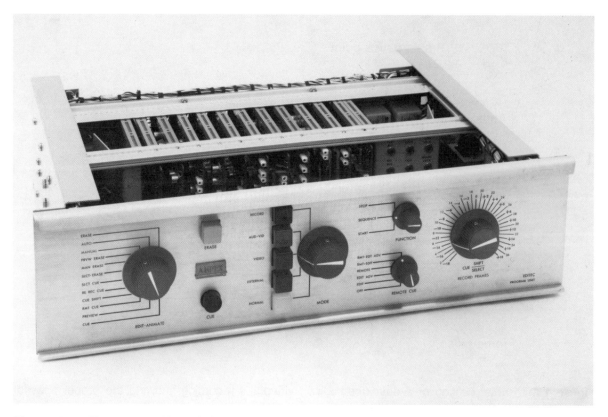

Figure 1.5: The Ampex EDITEC Control Unit used a short burst of audio tone from the VTR cue track to activate its editing features. Courtesy Ampex Corp.

tapes recorded during the Gemini and Apollo missions. Like the NASA system, the EECO code displayed elapsed time in hours, minutes and seconds. In addition, EECO added a device to track and display video frame numbers at the standard rate of 30 frames per second.

EECO also developed time code editing systems. As shown in Figure 1.6, the EECO 901 editing system featured thumb-wheel counters for entering the hour, minute, second and frame numbers of edit start and stop points. To enter the start and stop points, editors simply adjusted the thumb wheels for the appropriate counter displays.

The EECO system succeeded in making video editing a much more precise process. In fact, the EECO system became so popular it spawned a number of competitors—each of which used a different type of timing code. None of the codes were compatible, and the result was considerable confusion among video editors and technicians.

To remedy this problem, the Society of Motion Picture and Television Engineers (SMPTE) formed a commission in 1969 to develop an industry standard time code for video editing. Representatives from Ampex, EECO, Advertel and Central Dynamics Ltd. met, argued and, four years later, finally agreed on the standard we now call SMPTE time code. The SMPTE system was also adopted by the European Broadcasting Union (EBU), making SMPTE/EBU time code the international standard. (For a complete discussion of time code, see Chapter 2.)

Computerized Video Editing

Around the time that the industry was finally agreeing on a time code standard, a joint venture between CBS and Memorex Corp. produced the first computerized offline video editing system (a system used to create a workprint or edit decision list, rather than a finished video program). Called the CMX-600, it was a sophisticated set-up that featured large multiple-disk drives. Keying in on the videotape time code, the CMX-600 allowed instant access to any information (audio or video) recorded on the disk drive. The system featured a light pen, which editors used to search out and select edit points whose code numbers

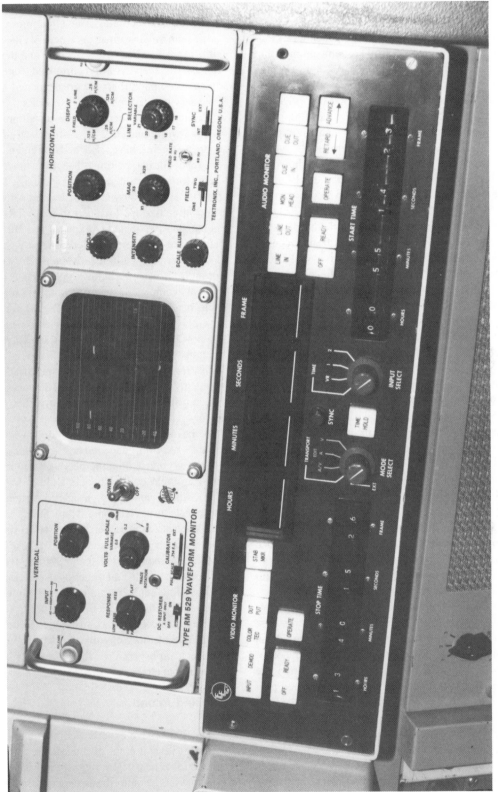

Figure 1.6: The EECO 901 "On Time" edit system (shown here mounted in an Ampex VR-2000 VTR) was the first broadcast-quality time code edit system. Photo by Darrell R. Anderson.

were displayed on the computer's CRT monitor, and an output device that generated edit decision lists on computer-readable punch tape. The punch tape could then be fed into the CMX-600's companion online system, the CMX EDIPRO 300.

Although the CMX-600 worked well, it suffered from one serious drawback—its $300,000 price tag. To many production facilities, this was simply too much to pay for an offline editing system. As a result, sales of the CMX-600 were relatively slow.

Sales of the CMX EDIPRO 300, however, flourished. Unlike its offline counterpart, the EDIPRO 300 could be used to edit finished video programs. Using SMPTE/EBU time code, the EDIPRO 300 gave editors frame-accurate control of both the playback and the record VTRs. It also allowed frame-accurate video switcher transitions, audio dissolves and a complete edit decision list readout from the system teletype. For this technical achievement, the CMX Company (now CMX Corp.) received an Emmy Award in 1973—along with the gratitude of professional video editors, who could finally give up their grease pencils.

Since 1973 CMX and other companies have produced several generations of newer and more sophisticated computer editing systems (see Figure 1.7). As the list of manufacturers in Appendix C indicates, there are at least 15 companies currently producing editing systems. This increase is due, in part, to the growing importance of video workprint (offline) editing.

Small-Format Editing

To this point, I have described editing systems and techniques that were designed for use with 2-inch broadcast VTRs. In recent years, broadcast and non-broadcast production studios have also become involved with small-format video editing—editing performed on 1-inch open-reel, 1/2-inch open-reel, 3/4-inch cassette, or 1/2-inch cassette VTRs. In broadcast studios, editing on small-format VTRs can provide an inexpensive means of generating workprints. In corporate facilities, small-format recorders are often the only machines available for editing the finished program—the videotape or cassette that serves as the final master copy.

In several respects, the history of small-format video editing parallels the history and development of large-format editing. When small-format video recorders first appeared in the 1960s, there was no reliable way to edit the tape. Although some editors tried mechanically cutting and splicing the tape, the slanted video

tracks used in small-format "helical-scan" video recording made accurate, clean edits impossible. However, with small-format open-reel VTRs, it was possible to mark the backs of the tape and then manually rewind the reels to a synchronized start point in a process comparable to the grease-pencil system used for years with 2-inch broadcast VTRs.

Earlier, I described how the development of time codes revolutionized 2-inch broadcast editing. Tape coding systems have also had a tremendous impact on small-format editing. The first practical use of small-format, offline editing is generally attributed to the late Hal Collins. Collins devised a technique for controlling two 1/2-inch reel-to-reel VTRs to produce an inexpensive, edited workprint. The 1/2-inch source tapes were copies of 2-inch broadcast tapes with time code superimposed in the TV picture. (See Chapter 5 for a discussion of window dupes.) Once offline editing was completed, the code numbers for edit points were hand-logged from the workprint for fast online assembly, using the original 2-inch broadcast-quality tapes. Collins used this method for years on shows such as "All in the Family," "Maude," and several Bob Hope and Milton Berle specials. In 1982 the Academy of Television Arts and Sciences recognized his contributions to video editing with a posthumous Emmy Award.

In the mid-1970s, "hands on" or "joystick" editing systems were introduced. These allowed small-format editors to control both the playback and record VTRs from a single control panel. Unlike the EECO and CMX editing controllers described earlier, these editing panels used electronic frame pulses in the video-tape control track, rather than SMPTE time code, to backspace the source and record tapes. Subsequent generations have added features that give the editor full control of all edit functions, plus optional time code editing capability, along with edit decision list readout, edit list management features, A-B roll editing interconnections to control video switching equipment, and memory circuitry for storing and recalling edit decision lists (all of which are described in detail in later chapters.)

Looking Ahead

The 1980s have experienced a deluge of new products for several areas of the video editing and post-production process. The biggest emphasis in the broadcast area has been in film-to-video offline editing systems, digital video effects systems and electronic animation and graphics systems. Random-access laser

Figure 1.7: The CMX 50 offline cassette edit system, introduced in 1974. Courtesy of CMX Corp.

disc or VCR-based film-to-video edit systems, such as the Montage, Editflex and CMX-6000, are convincing increasing numbers of filmmakers to forego the traditional methods of film editing in favor of the new electronic methods that are proving to be faster and more economical.

Random access editing is not a new phenomenon. As I previously described, the CMX-600 was an instant, and random access, system using large and bulky multiple-disc drives (the only appropriate storage medium available in the early 1970s). Today's random access systems use either multiple banks of 1/2-inch VCR playbacks or the new, and relatively expensive, laser disc format. Random accessibility to the 1/2-inch VCR systems is made possible by having multiple copies of the same source material simultaneously loaded into the VCR machines. No recording actually takes place; instead, a continuous real-time preview reproduces the edited material according to the edit decision list. This permits internal editing changes without recording down a generation as in traditional linear video editing. The same philosophy holds true for laser disc systems except that laser discs feature almost instant cueing to any segment on the disc, therefore requiring fewer multiple copies.

Laser disc-based linear offline video edit systems,

low-cost personal computers functioning as edit controllers and edit list management tools (such as those discussed in Chapter 5) are just beginning to make their bid as the post-production tools of the future.

The field of digital video effects has exploded to the point where the small corporate or educational video professional can now afford one of the basic "digital" effects devices such as the FOR.A System described in Chapter 7. Over on the high-end of the product scale, effects systems such as Quantel's Mirage permit images to be programmed virtually to any shape that the user can image. This proliferation of effects devices for all budgets was brought on by the immense popularity of high-tech effects, the expanding video industry base and ongoing achievements in digital technology.

The popularity of high-tech effects has also led to a tremendous growth in the electronic graphics industry. Advancing from the virtual video typewriter of less than 10 years ago, today's advanced graphics systems not only create lettering in innumerable fonts, sizes and colors, but they also compose animated images, "paint" video frames and create intricate scenes of video artwork.

The year 1986 marked the thirtieth anniversary of the first videotape recorder. Quite a lot has happened since that first "on air" operation from Television City in Hollywood. Even those engineers at Ampex could never have foreseen the enormity of what would transpire, yet their invention spawned the post-production industry as we know it today.

To give you an idea, in less than a third of a century the post-production industry has developed into an over $4 billion per year business involving a conservatively estimated 5000 companies. These include TV stations, corporate and industrial post-production houses and audio sweetening facilities. This does not even begin to count the scores of equipment manufacturers. A survey conducted in late 1986 by the Harry Heller Research Corp. revealed that corporate and industrial video professionals, along with commercial work, account for more that 60% of the entire post-production business. Television programming, the "glitter and glamour" area, represented only 17%, while ENG accounted for 5% and feature films, another high profile area, represented only 6%. The remaining percentage of "other" business included animation graphics and effects, CCTV, music videos, home videos, interactive videos and various consumer projects.

It remains to be seen if this almost unprecedented rate of growth will stabilize or continue surging ahead as new innovations in post-production technology continue to surface.

As this brief history suggests, video editing has come a long way in a short time. In fact, by the time you are reading this book, there will undoubtedly be newer, more sophisticated editing systems available that will offer more features for less money than the equipment I have described.

In the later chapters on editing techniques and processes, I have tried to focus on the basic principles and procedures that are fundamental to effective post-production. As a result, most of the information contained in those chapters should remain valid, regardless of which editing systems and accessories come to reign as "state-of-the-art" over the next few years. After all, good preparation is the key to good results, and good preparation requires a solid foundation in basic post-production principles and practices—including the fundamental artistic principles that have guided film and video editors for years.

THE ART AND TECHNIQUE OF VIDEO EDITING

Many of the artistic principles used in video editing developed from tricks and techniques perfected by filmmakers. Sergei Eisenstein, Alfred Hitchcock and other great film directors have become fixtures in editing history for their unique methods of framing and sequencing shots and for their innovative pacing and transitional techniques. Like their film counterparts, video directors and editors face three fundamental post-production decisions. They must select the shots they will include in the edited program; they must decide on the sequence of shots that best conveys the message they are trying to communicate; and they must determine how to time and pace the shots to create the exact effect they are after.

The amount of control a video editor exercises over these decisions differs from production to production. In dramatic productions, video editors often work from a script marked with the producer's or director's instructions. When this is the case, the editor usually has very little control over the selection and sequencing of shots. On the other hand, editors of documentary-style productions frequently find themselves working from a script or scenario that is much less detailed. As a result, documentary editors often play a key role in selecting shots and determining edit points and transitions.

Of course, the degree of creative control that an editor is asked to assume in a production also depends on his working relationship with the executives in charge of the program. Producers and directors tend to delegate the greatest degree of authority to editors whom they know and trust, and who have proven that they are capable of making decisions about shot selection and sequencing. As a result, editors who hope to become creative participants in the post-production process need to know much more than how to operate editing equipment. To make the right creative decisions, they also need a fundamental working knowledge of camera shots, camera angles and the principles of pacing and program continuity.

SHOTS AND ANGLES

In most post-production sessions, editors compile the final edited program from raw production footage recorded on videotape workprints. Usually, the raw footage includes many more shots than the editor will finally use and many scenes that were shot from several different angles and perspectives. Since each production is unique, there is no single set of guidelines that editors can use to determine which are the "right" shots. There are, however, a number of basic definitions and principles that can help editors determine

Wide shot

Medium shot

Close-up

Figure 1.8: The three most frequently used camera shots. Photos © copyright 1983, Embassy Television.

which camera shots and angles are right for a particular scene or sequence.

Types of Shots

Camera shots are categorized by the image size—the size of the subject matter in relation to the total area included in the shot. Since each type of shot communicates a different message to viewers, each serves a different purpose in a finished video program. I have listed the primary types of shots used in video production below, along with some suggestions for how each could be used in an edited program. The three most frequently used shots are illustrated in Figure 1.8.

Extreme Wide Shots (EWS)

Extreme wide shots are panoramic long-distance shots that take in the widest possible view of an area. An EWS is usually used as an establishing shot—a shot that establishes the setting at the beginning of a scene or sequence. In an EWS, the scene itself serves as the subject matter. With the camera set at such a wide angle, it is impossible to make out individual faces or to focus on specific details of the scene. Since television, with current technology, is essentially a small-screen medium, an EWS included in a TV program will not usually have the same strong impact as an EWS in a feature film. Recognizing the limitations of the small screen, many TV producers prefer to use wide shots, rather than extreme wide shots, to establish a scene or setting.

Wide Shots (WS)

In wide shots, the camera is closer to the scene than it is in extreme wide shots, allowing viewers to recognize distinguishing details and features. For example, a wide shot that opens a situation comedy might show the full view of the house or room where the action is set. If the scene already includes actors, the WS would allow viewers to see where the characters are standing in relation to each other. Once the WS has set the stage, the director or editor would probably move to a medium shot that shows the characters engaged in dialogue.

Basically, then, wide shots are used at the head of a scene or sequence to help avoid the confusion viewers would feel if performers entered and moved about a scene shown only in medium shots or close-ups. In addition, wide shots are sometimes used in the middle of scenes, to reestablish the location of characters in sequences where the performers move about a great deal.

Medium Shots (MS)

As shown in Figure 1.8, medium shots usually show characters from the waist up. As mentioned above, many TV directors start a sequence with a wide shot establishing the scene, then cut to an MS showing the primary characters. This brings viewers closer to the action without the jarring effect that is sometimes created when a program moves from a WS directly to a close-up.

Medium shots can also be used to follow characters as they move around a scene. In fact, following character movement with an MS can be more effective and less jarring than continually cutting from an MS to a WS and back again, as long as the director has been careful to compose the medium shots correctly.

The most common MS in TV production is probably the ''medium two-shot''—a medium shot showing two performers. The medium two-shot may show both performers in profile, or it may feature an angled view in which the audience sees the scene from the perspective of one of the performers.

Generally, I use two-shots showing both performers in profile (''profile two-shots'') when a third character is about to enter the frame between the two principal performers, or when some other important action will take place between them. Otherwise, I try to avoid extended or extensive use of the profile two-shot since it tends to hide the characters' facial expressions.

In angled two-shots (or ''cross-twos''), the camera cuts between shots that show the scene from the perspective of each individual performer. This is an especially effective technique in TV dramas and situation comedies since it allows viewers to follow dialogue and the reactions of performers. In cutting between shots, directors and editors should make sure that both shots feature the same image size and comparable camera angles.

Close-Ups (CU)

In video production, a close-up can range from a shot showing a single character from the bust up (a ''bust shot'') to a shot showing only the character's face (an ''extreme close-up''). Since TV is a small screen medium, close-ups can be one of the most useful and effective shots in an editor's arsenal, provided they are properly framed and recorded during the production session.

High-angle shot

Low-angle shot

Normal (straight) angle

Figure 1.9: The three basic camera angles. Photos © copyright 1983, Embassy Television.

Close-ups are most often used within scenes to focus viewers' attention on a performer's facial expressions or to highlight a particular portion of the scene. However, in an intentional reversal of this convention, some directors open scenes with close-up shots and then gradually pull the camera back to a medium or wide shot. When used carefully, this technique of gradually revealing the setting and circumstances can add suspense to a scene, especially when the opening CU shows the performer expressing strong emotion (surprise, fear, anger, etc.). Close-ups can also be used to add dramatic or comedic impact, to magnify an important object and to provide cutaway or transitional images. In addition, close-ups taken from the point of view of a person performing a task are invaluable in industrial training tapes.

Playing the Angles

Anyone who has taken an introductory course in film or TV production is familiar with the three basic camera angles: high angle, low angle and normal, or straight, angle (see Figure 1.9). In high angle shots, the camera looks down on the performers from above, giving viewers a sense of power and command over the scene. In low-angle shots, the camera is placed below the performers, looking up, making the performers seem larger and more dominant. Finally, as its name suggests, the normal angle shot shows the scene straight on, putting the performers and the audience on the same level. Three versions of these basic angles are the objective angle, the subjective angle and the point-of-view angle.

Objective Angle

The objective angle is the most popular camera angle used in video production. As its name suggests, the objective angle does not present the scene from the perspective of any one character. Instead, it places the camera outside the scene looking in and places viewers in the position of unseen observers. As a result, when editors and directors are using the objective angle, they should make sure that no shot shows a performer looking directly into the camera. For viewers who have been watching the scene as unseen observers, such direct eye-to-eye contact with a performer can be very disconcerting.

Subjective Angle

As its name suggests, the subjective angle places viewers in the scene so they see events from the perspective of an on-screen performer. For example, in a film or TV show that depicts a World War II bombing raid, the director might start a sequence with a normal angle medium shot showing a bombardier peering into his bombsight. Then, to show viewers the target area as the bombardier sees it, the director might cut to a subjective angle shot in which the camera actually looks through the bombsight. In other words, in a subjective angle shot, the camera acts as a stand-in for a performer, allowing the audience to see the scene as the performer would see it. The subjective angle is also very useful in industrial training films, to show a manufacturing process as an employee performing the process would see it.

Point-of-View Angle

The point-of-view angle combines features of both objective and subjective angle shots. In the point-of-view angle, the camera is usually placed next to a performer. In other words, viewers do not see the action directly through the performer's eyes as they would in the subjective angle. Instead, viewers watch from alongside the performer, like an invisible observer standing next to the character.

Any shot can become a point-of-view shot, as long as it is preceded by a shot showing a performer reacting to something off-camera. For example, an industrial training film might begin with a medium shot showing a plant manager explaining a manufacturing process, then turning to look off-screen. If the next shot shows another part of the plant from the manager's angle, it's a point-of-view shot—regardless of whether it's a wide shot, a medium shot or a close-up.

Point-of-view shots are also very popular in situation comedies, where they are often used to follow dialogue and action from a performer's perspective. Notice, however, that the actors never look directly into the camera (as they might in a subjective angle). As a result, the point-of-view angle gives viewers the intimacy of being next to the performer while still maintaining the appearance of objectivity.

CONSTRUCTING CONTINUITY

Once he* has a grasp on the various shots and angles he has to work with, the editor can get to work putting

---

*—or she; "he" is used throughout this book only for simplicity and is intended to include women as well.

all the pieces together. Regardless of whether he is working from a ''camera blocked script'' (a script that includes exact instructions for where each shot should be placed) or from the director's rough notes, an editor almost always pieces the program together in two separate stages: rough cutting and fine cutting.

Rough Cutting

In the rough-cut stage, the editor's goal is to transform shots selected from the raw production footage into a continuous story with smooth cutting between shots. Once it is edited, the rough-cut reel allows the director and the producer to get an overall look at the story and to evaluate its strong and weak points. The pacing may be too slow or too quick in spots; the director may recall better shots than those included in the rough-cut reel; or the director may decide to rearrange the sequence of several shots or segments.

Fine Cutting

During fine cutting, the editor refines the rough cut into a smoothly flowing story, paying particular attention to the pacing and performance changes suggested by the director and producer. Fine cutting can be accomplished by redoing the original edits on the master reel or by using the rough cut as another production reel and recording down a generation.

In actual practice, professional video editors usually combine the features of rough cutting and fine cutting. For example, many rough-cut reels contain video transitions and audio mixings, features normally associated with fine-cut editing.

Continuity Cutting

Notice that I have emphasized the importance of creating *smooth* video cuts, that is, cuts that do not create a noticeable distraction for the viewer. In other words, a smoothly edited program maintains the illusion that the story is being told in a continuous, uninterrupted manner.

The primary requirement of smooth cutting is that the scene action flows logically across consecutive shots. This is important for both background visuals and principal character movements. The history of video editing is filled with horror stories about finished programs that included drastic lighting changes between scenes, primary props that were not reset between takes and actresses' hair styles that changed during the course of a scene. My own favorite story occurred during the three-day shoot of a dramatic pro-

Figure 1.10:   A neutral direction shot can smooth a change of screen direction.

gram. As they were reviewing rough-cut reels from the shoot, the editor and director discovered that one of the principal actors had worn two different colored suits during the various stages of shooting—a mistake that created a drastic break in continuity.

Maintaining Proper Screen Direction

To sustain continuity, an editor must maintain proper screen direction. For example, a subject traveling off-camera from left to right should not be picked up in the next shot traveling from right to left. If an editor makes this mistake, the viewer will become confused and assume the subject has turned around and is returning. In addition, once screen direction is established, it must be maintained throughout the sequence of shots involved with that particular movement.

Of course, it is sometimes necessary to change screen direction between shots. One way of doing this smoothly is to bridge the two shots with a shot showing neutral screen direction. The neutral image can show the character or characters either coming straight toward or moving directly away from the camera. Since this type of shot appears nondirectional, it smooths the transition for viewers (see Figure 1.10). The neutral direction shot can also be used to shock viewers, especially when dramatic or suspenseful action suddenly shifts direction and comes straight toward the viewer.

Screen Position

To sustain continuity, it is also important to maintain performers' screen position in consecutive shots. Failure to do so distracts viewers since it creates unreal situations in which performers appear to jump instantaneously to different screen locations. If possible, screen position problems should be avoided during production through careful shot composition. However, screen position problems can also appear during the editing stage, particularly when edits involve cutting between shots of performers in motion. For example, if a character who is framed at center screen turns and moves to the right, cutting to a shot with the same performer now framed on the left side of the screen will give the viewer the impression that the performer has jumped backwards. This is even more of a problem when an editor is cutting a sequence involving three performers in which consecutive shots show two of the performers from corresponding angles. If one shot shows the center performer and the performer on the right and the next shot shows the center performer and the performer on the left, the center character will

seem to jump from the left side to the right side of the screen. In this situation, editors can only hope that the director has also recorded shots of the individual performers that can be intercut between the two shots.

To avoid disruptive discontinuity, editors should also make sure that consecutive shots showing the same performers are taken from distinctly different angles. Consecutive shots of the same performers taken from relatively small angle differences will give the viewer the impression that the characters are staying still and the set is moving. Again, this is a problem that is best dealt with during production, since it is very difficult to change the composition of shots once a program enters the post-production stage. Although digital effects devices (see Chapters 3 and 7) can sometimes enlarge or reposition the image, the shot will suffer resolution degradation if this is done to excess.

Matching Action

Matching the actions of the characters through consecutive shots may be the most fundamental requirement of smooth continuity cutting. For example, think of a scene in which a character approaches a door, opens it and enters a room. To cut this sequence properly, the editor must avoid duplication of action across the cuts. Also, there should be no jump in the action, as would be the case if the editor cut from the shot in which the door starts to open to a shot of a fully opened door. Matching character action is a comparatively simple matter as long as the performers do their part. Experienced actors and actresses know how to repeat the timing of their actions and their dialogue. However, less experienced actors frequently make it extremely hard for an editor to piece together this type of scene, since novice actors often find it difficult to match action and dialogue across shots.

Matching Action Cuts

This brings us to another aspect of matching action: matching action cuts. Few editors agree on the best method for cutting action. Put eight editors into a room and describe an action scene, and you will get eight different opinions on where and how to edit it.

Take, for example, a scene showing a man sitting in a chair. The man drinks from a glass, sets the glass down, rises to his feet and exits. Obviously, there are many ways to cut this sequence. A director may want to begin with a close-up showing the expression on the man's face as he drinks, followed by a medium shot showing the man putting the glass down, followed by a wide shot showing him rising and exiting.

Where should the editor make the cuts between shots? He could cut to the medium shot before the man finishes drinking, but that would telegraph the action to come. On the other hand, cutting after the man starts to put the glass down would show the movement, but it would not make the best use of the action. The third and possibly best cutting point is the moment when the man begins his move to put the glass down. By cutting at this point, the move would be initiated in the close-up and carried across the cut to the medium shot, enhancing the motion by cutting at the precise moment the scene changes from rest to action.

The same holds true for the man's rise and exit. When the man starts his forward movement (the point of change from rest to action), the editor could cut to the wide shot. When done consistently, this type of action cutting will create the illusion of a smooth, uninterrupted action sequence for the viewer.

PACE AND TIMING

The ability to lengthen or shorten shots and sequences is one of the editor's most powerful tools. By shortening or drawing out a scene, an editor can often alter the pacing and impact of a sequence. For example, at the request of a director or producer, I've often adjusted edit points to transform a short kiss between two performers into a lingering, more passionate kiss.

Of course, there are other, less romantic reasons for altering pacing. For example, to make a performer move across stage more quickly, an editor might cut from a wide shot of the performer to a tighter, more neutral angle. This would allow the editor to save needed time by trimming two to three steps out of the sequence without creating any noticeable discontinuity.

When they alter a scene's natural pacing, editors should make sure that the new pacing doesn't introduce an erratic texture to the flow of the sequence. For example, when an editor is cutting dialogue sequences that include several reaction shots of each player, changing the length of one reaction shot might necessitate changing others as well to maintain a consistent flow.

Probably the most basic example of altering pacing for audience impact is the dramatic chase sequence in which the editor cuts between the hero and the villain at a progressively quicker rate as a way of heightening the exciting effect of the action. Timing shots for

laughs in a TV comedy sketch is another example. Most TV comedy is played at a slick pace. Unless the performance was recorded entirely before a live audience, editors often have trouble determining how much space they should leave between shots for laughter. On the one hand, following a joke with too large a laugh gap creates an unnatural pause that can slow the program's pace. On the other hand, a gap that's too short may mean that the laughter obscures the dialogue that follows. On balance, it is usually better to keep the laugh breaks on the short side, so the pacing stays intact, rather than to create a huge gap for a joke that doesn't deserve it.

Montage Sequences

Montage sequences provide a particularly interesting example of the relationship between editing and pacing. In video or film montage, a number of different images are combined to create a strong, often abstract impression. For example, rock videos (video clips originally developed as record album promotions) are often built around montage sequences that include shots of the musicians in concert, action scenes that relate to the lyrics of the song, or both. In this type of musical montage, the pace and length of the shots are usually determined by the tempo of the song, and the entire sequence is usually pieced together in post-production.

In documentary or dramatic productions, montage sequences are sometimes intermixed with more conventional narrative segments. When this is the case, the pace of the montage sequence is often shaped by the pace of the surrounding segments or by the particular production needs that originally prompted the producers to use the montage material. For example, in a TV documentary on the student protest movement of the 1960s, a producer might instruct an editor to intermix interviews and profiles of the movement's leaders with fast-paced montage sequences of motion and still footage taken at demonstrations and rallies. When they intermix narrative and montage sequences in this way, editors should be careful not to let the montage sequences run too long. In addition, adding music in keeping with the mood of the montage can often help strengthen the impact of the sequences.

Audio and Other Concerns

The program's audio track can also play a key role in smoothing the transition between shots. For example, in continuous action sequences, it is often desirable to let the audio flow uninterrupted across the edit, either to cue a character's attention before the edit point or to help smooth over a mismatched action edit. Overlapping audio can be equally effective across static shots in dialogue sequences since it will add dimension and smoothness to an otherwise actionless string of shots.

To ensure a smooth transition across cuts, the editor must also pay close attention to audio levels and equalization. Viewers will perceive unintended audio level changes from one shot to the next as a disturbing, distracting irritation. Editors should also be sure to maintain the luminance and color values of the shot sequences across edits. In fact, drastic light and color value changes occurring between supposedly continuous action sequences can create a jarring visual effect every bit as annoying as a mismatched action edit. In most cases, these light and color mismatches can easily be corrected by adjusting the appropriate level controls on the playback VTR. However, with particularly troublesome shots, it may be necessary to run the video through color correction devices during the on-line editing process.

CONCLUSION

In the final analysis, there are several ways to edit almost any particular sequence. In this chapter, we have only touched on some of the more fundamental considerations and concerns that shape editing decisions and the final look and feel of the edited material. As mentioned earlier, it's always best to anticipate problems during the production stage. Editors are not magicians; they can only assemble the project with the footage provided. If the right material isn't in the can after production, don't expect to find it during post-production. A friend of mine refers to this as the ''fast pig'' theory. If a producer brings in a poorly produced and shot project (i.e., a pig), he shouldn't expect an editor to turn it into a thoroughbred race horse. All the editor can do is give the producer back a faster pig.

2 The TV Signal, Time Codes and Videotape Formats

Many aspiring editors seem to believe that they will automatically become skilled video editors as soon as they learn to operate basic editing equipment. This is about as realistic as believing that a person who learns to type will suddenly be capable of writing a classic novel. A professional in any field requires far more than a simple working knowledge of the most basic tools and machines of the trade. A person certainly isn't considered a professional film editor simply because he's learned to glue two strips of celluloid together.

The same holds true for professional video editors. Video editing is both a technical and a creative process. To become a true professional, then, an editor must become skilled and knowledgeable in both aspects. He must become familiar with the artistic principles and conventions of his craft, and he must acquire a thorough understanding of the technical processes and principles that determine the quality of the finished, edited program. In Chapter 1, I reviewed several of the fundamental artistic principles behind the editor's craft. In this chapter, I will focus on the technical end of the profession—the basic technical principles and procedures that govern the operation of today's sophisticated editing systems.

Before I scare any readers away, I should pause to say that you do not need an engineering degree to qualify as a professional video editor. However, there are a great many pitfalls for those who choose to ignore or neglect the technical aspect of editing. I am reminded of a colleague who paid little attention to the technical quality of the video signal on the edited master, only to have to explain to the producer why the program couldn't be broadcast until it was totally re-edited. To him, the signals displayed on the monitoring equipment in the editing bay amounted to little more than "strange patterns." He could have prevented this problem if he had only taken the time to discover what the patterns meant.

In this chapter, you will learn about one source of those "strange patterns"—the color television signal. We will also look at the various types of videotape time code and the different video recording formats. In writing this technical primer, I have intentionally *not* included any of the complicated engineering equations or mathematical measurements that are sometimes used by video technicians. For readers who are interested, those topics are covered thoroughly in books that approach video from an engineering perspective. Several are included in the bibliography of this book.

TELEVISION FIELDS AND INTERLACED SCANNING

In the United States, the standard television signal is created by an electronic beam scanning a picture tube at a rate of 525 lines of picture information 30 times each second. As shown in Figure 2.1, the scanning beam starts at the upper left corner of the picture tube raster and moves from left to right, dropping down a line and then retracing when it reaches the

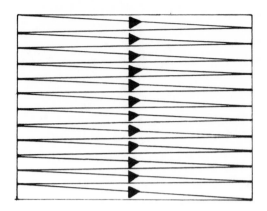

Figure 2.1: Original television signal scanning method.

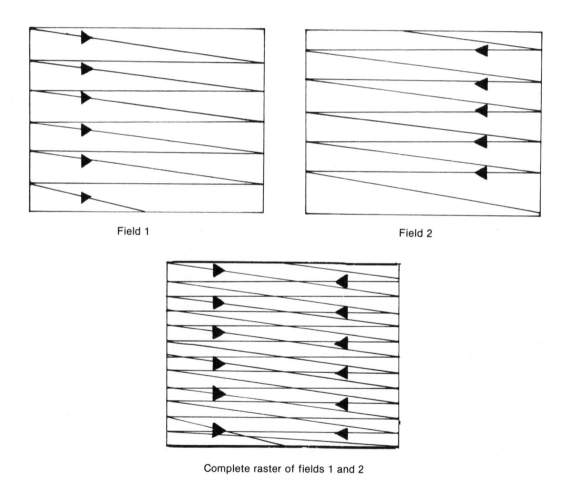

Field 1

Field 2

Complete raster of fields 1 and 2

Figure 2.2: Interlaced scanning.

right-hand edge of the picture tube, or raster, area. But there is an odd twist to this system. Early in the development of television, engineers discovered that, although 30 scans per second produced a picture on the television monitor, the picture contained considerable flicker. To resolve this problem, the designers first tried increasing the scan rate to 60 scans per second. Unfortunately, although this remedy did reduce the flicker, it introduced other problems when the signal was transmitted.

To minimize both sets of problems, television's engineering forefathers developed a system called "interlaced scanning." Interlaced scanning works on the principle of dividing the television frame into two separate fields, each containing 262.5 (half of the original 525) lines of picture information. Field one contains the odd numbered lines (1, 3, 5, 7, etc.), and field two contains the even numbered lines (2, 4, 6, 8, etc.). Field one is scanned first, producing an image containing only the odd-numbered lines. Then, under the control of vertical synchronization pulses, the electron beam is returned to the top center of the raster, where it begins scanning the second field, the even-numbered lines. By the time the beam reaches the bottom right corner of the raster, it has filled in the rest of the picture (see Figure 2.2). In other words, although the standard U.S. television signal creates a TV picture at the rate of 30 frames per second, each frame contains two interlaced fields, resulting in an actual scanning rate of 60 fields per second.

THE COMPOSITE VIDEO SIGNAL

The NTSC (National Television Standards Committee) television signal is a complex combination of pulses and sine waves that carry seven distinct types of information necessary to create a television image. Together, the seven types of information form what is known as the television composite waveform, or composite video for short.

As shown in Figure 2.3, the composite video signal includes:

- Horizontial line synchronizing pulses
- Color reference burst
- Reference black level
- Picture luminance information
- Color saturation information
- Color hue information

An additional part of the composite video signal, the vertical synchronizing pulse, does not show in this illustration. It will be discussed in the section on the vertical blanking interval.

Horizontal Line Synchronizing Pulses

Horizontal line synchronizing pulses (or horizontal sync) lock the scanning electron beam of the TV monitor so that each line of picture information will start at the same lateral position during the scanning process. The pulses occur before each horizontal line of picture information.

Color Reference Burst

Color reference burst is a 3.58 MHz (megahertz) sine wave (usually of nine cycles duration) added just before the picture information on each horizontal line of video. An expanded view of color reference burst is shown in Figure 2.4. Color reference burst is the standard reference against which both hue and color saturation information are measured. This assures that the color information displayed by the monitor is the same as that picked up by the camera.

Reference Black Level

Also known as "pedestal" or "set-up," reference black level is defined as 7.5 IRE (Institute of Radio Engineers) units of picture amplitude for color TV and 10 IRE units of picture amplitude for black-and-white TV (140 IRE units equal one volt). Many video technicians prefer to describe reference black level as a percentage of total picture amplitude—7.5% for color and 10% for black-and-white.

Picture Luminance Information

Picture luminance refers to the brightness levels of the image scanned. The picture luminance information falls within a gray scale range of amplitudes from reference black (at 7.5 IRE units) to peak white (at 100 IRE units).

Color Saturation

Impressed upon the picture luminance information is a 3.58 MHz sine wave frequency (called the 3.58 MHz subcarrier) which, when its amplitude is compared with the 3.58 MHz color reference burst, determines the color saturation of the displayed picture. The greater the amplitude, the greater the saturation (and the deeper the color).

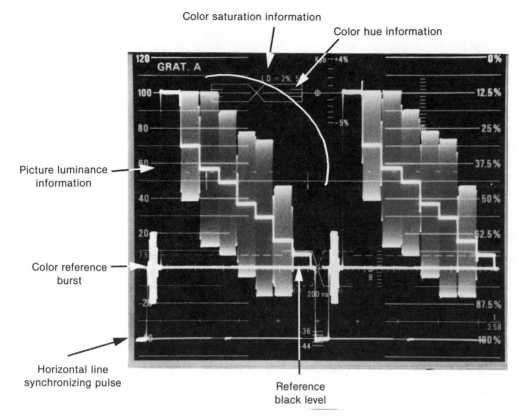

Figure 2.3: Two horizontal scan lines of the seven-bar NTSC color bar test signal showing the components of the video signal. Photo courtesy Tektronix, Inc.

Color Hue

Color hue, the other element present in the 3.58 MHz subcarrier, is determined by the phase of the subcarrier when compared to the color reference burst signal. The phase difference determines the hue of the color displayed on the television monitor.

BLANKING INTERVALS

There are two other important elements contained in the composite video signal: the horizontal blanking interval and the vertical blanking interval.

Horizontal Blanking Interval

The horizontal blanking interval (shown in Figure 2.5) is defined as the section of the composite video signal from the trailing edge of picture information on one horizontal video line to the leading edge of picture information on the following horizontal video line. The horizontal blanking interval occurs when, under the direction of the horizontal line synchronizing pulses, the scanning beam returns to the left side of

the picture tube to begin scanning the next line of video information. Contained in this interval are the front porch, the breezeway and the back porch, as well as the previously described horizontal sync pulse and color reference burst.

Front Porch

The front porch extends from the end of the picture information to the leading edge of the next horizontal sync pulse. The primary function of the front porch is to blank the receiver just prior to horizontal retrace (that period when the electron beam is returned to the left side of the raster to begin another horizontal picture line).

Back Porch

The back porch extends from the trailing edge of the horizontal sync pulse to the start of picture information. Its primary purpose is to keep the receiver blanked until the start of active picture information. In addition, the color reference burst signal is inserted during this period.

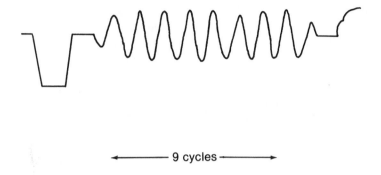

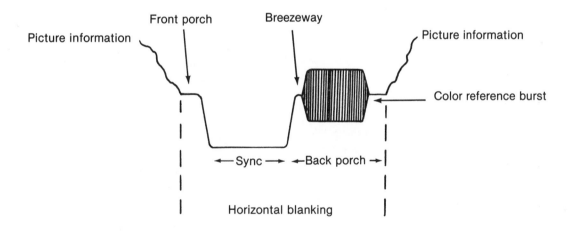

Figure 2.4: Expanded wave pattern of the 3.58 MHz color reference burst.

Figure 2.5: Expanded view of the horizontal blanking interval.

Breezeway

The breezeway is that part of the back porch extending from the trailing edge of the horizontal sync pulse to the start of the color reference burst signal. Its purpose is to maintain receiver blanking when the signal is coming out of the horizontal retrace period.

Vertical Blanking Interval

The vertical blanking interval, or VBI (shown in Figure 2.6), is the portion of the composite video signal from the end of the last horizontal line containing picture information in one television field to the point starting the first horizontal line containing picture information in the following television field. Controlled by the vertical synchronization (vertical sync) pulses, the VBI is the period during which the TV screen goes blank as the scanning beam returns to the top of the screen to begin scanning the next video field.

The VBI typically contains 20 horizontal lines of blanking, the most important of which are the first nine. These nine lines contain the pre-equalizing pulses, the vertical sync pulses and the postequalizing pulses.

Pre-equalizing and Postequalizing Pulses

There are six equalizing pulses of half a horizontal line duration each, preceding and following the vertical synchronization pulses. The equalizing pulses assure proper interlace synchronization during the vertical synchronization period.

Vertical Synchronizing Pulses

Vertical synchronizing pulses cue the electron scanning beam to the correct start point to begin scanning the next television field. In 1953, when the NTSC color television signal was being devised, it was found that mixing the 60 Hz vertical synchronizing pulse rate with the 3.58 MHz color reference subcarrier frequency during transmission produced visible interference. (The sound frequencies became mixed with picture information in home television receivers.) To alleviate this problem, the vertical synchronizing rate was changed slightly to 59.94 Hz. This small change eliminated the interference and still allowed 60 Hz television receivers to function properly. As a result, all NTSC color television equipment designed on the U.S. standard generates video and timing pulse signals at a 59.94 Hz synchronizing rate.

Other horizontal lines within the VBI have been used for other purposes. For example, lines 10 through 16 have served as a slot for the vertical interval time code used with some helical editing systems that feature slow-motion and freeze-frame techniques. (SMPTE/EBU vertical interval time code is discussed later in this chapter.) In addition, horizontal lines 17 through 20 may be used for inserting test and alignment signals. The most recent innovation in vertical interval signal applications is closed captioning code for the hearing impaired.

FCC Blanking Requirements

According to current FCC rules, the horizontal blanking interval should not exceed 11.4 micro-seconds during transmission. To make sure they comply with this regulation, production companies typically shoot material with the horizontal blanking set at 10.7 microseconds. This safety factor allows for the inevitable widening of the horizontal blanking interval that occurs during post-production and transmission.

FCC requirements state that the vertical blanking interval should fall in the range of 19 to 21 horizontal lines. Since the VBI generally remains constant from production through transmission, there is no need to build in a safety factor.

THE ELECTRONIC VIDEO SIGNAL AND THE COLOR TV PICTURE

Now that you have received a very basic introduction to TV signal theory, how do you use your theoretical knowledge to analyze the picture you actually see on the television monitor? To answer that question, let's try applying our basic signal theory to a color television reference signal that is familiar to anyone working in the television industry: the NTSC split field color bars. (A schematic representation of the NTSC split field color bars is included in Figure 2.7.)

NTSC Split Field Color Bar Reference Signal

The name "split field" does not refer to the television signal itself. Instead, it refers to the point three-quarters of the way down the monitor presentation where the signal changes from the display of primary and complementary colors to a display of the $-I$ and $+Q$ test signals, the 100% reference white level and the 7.5% reference black level.

The three primary colors used in the NTSC system (the colors that are combined to reproduce all other colors) are red, green and blue. The complementary colors include yellow, cyan and magenta. Red, green

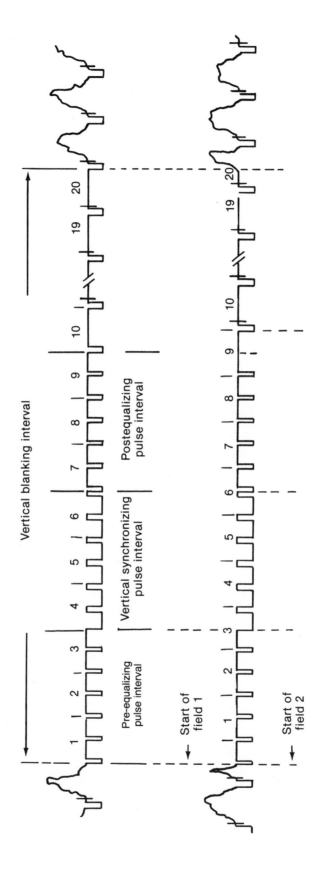

Figure 2.6:   Expanded view of the NTSC vertical blanking interval.

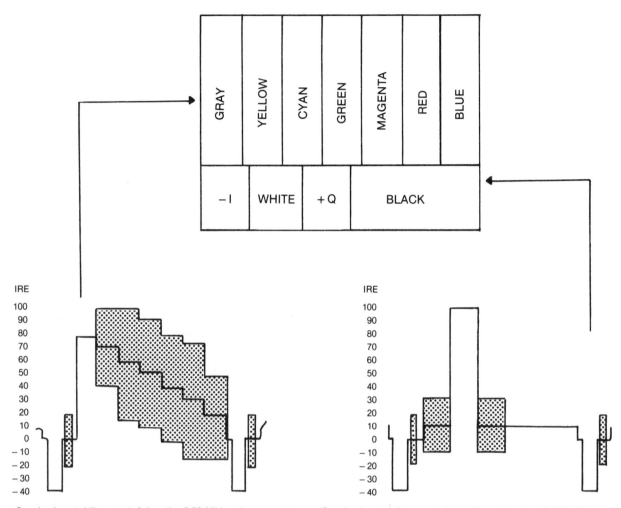

One horizontal line containing the 3.58 MHz primary and complementary color subcarrier frequency impressed on the luminance gray scale.

One horizontal line containing the − I and + Q 3.58 MHz color subcarrier frequency impressed on the luminance reference black level and peak white level.

Figure 2.7: Horizontal scan lines for the NTSC split field color bar reference signal.

and blue (commonly referred to as RGB) were chosen as the primary colors in 1953, mainly because of the light output characteristics of the phosphors used in color television picture tubes, and because they were primary colors that permitted a wide range of color reproduction.

Figure 2.7 illustrates one horizontal scan line of the upper portion of the signal, containing the primary and complementary colors, and one horizontal scan line of the lower portion of the signal. For the former, the shaded areas indicate the 3.58 MHz subcarrier impressed upon a graduating gray scale series of steps. For instance, the gray step (far left) has no subcarrier impressed upon it, but equals 77 units (or 77%) of video level. The yellow color (second from left) has a gray scale brightness of 69 units (or 69%) and a

subcarrier with a lower limit equaling 38 units and an upper limit equaling 100 units.

Note that the color reference burst equals 40 total units of video level (+ 20 to − 20). This level must be strictly adhered to since the color reference burst is the primary signal factor against which all colors are compared.

Figure 2.7 also illustrates one horizontal line of the − I and + Q signals, reference white level and black level. The − I and + Q signals were developed from the results of color perception testing conducted by NTSC in the late 1940s and early 1950s. From those tests, NTSC determined that color differences in the yellow-to-orange region were readily discernible, while color differences in the green-to-cyan region were perceived as varying shades of gray. Through a series of complex mathematical calculations (which there is really no need to introduce here), NTSC determined that the −I and +Q signals, in their current configuration, would serve as the reference for those areas of maximum and minimum color difference perception. As noted earlier, reference white equals 100 IRE units (100%) and reference black equals 7.5 IRE units (7.5%)

Figure 2.8 is a photo of the split field color bars taken from the face of a television waveform monitor displaying the TV signal at a two horizontal line rate. Notice that the presentation appears as a superimposition of the separate scan lines in Figure 2.7. This is the presentation generally viewed by video editors since both waveforms can be seen at one time.

Figure 2.9 is a photo of the split field color bars taken from the same television waveform monitor as it displays both scanning fields or one full frame. The waveform is read from left to right, with the left side

of the screen representing the top of the monitor picture. A comparison of Figure 2.9 with the drawing of color bars in Figure 2.7 shows that, as you look toward the right, you are actually moving down the monitor picture. The point at which the amplitude appears to drop off denotes the point on the monitor picture where the signal changes from primary and complementary colors to the −I and the +Q test signals with reference white and reference black. Notice that the 100% level—the reference white bar—is bright through this area. If Figure 2.9 were greatly expanded, the first three-fourths of the presentation would look like the first scan line of Figure 2.7, while the remaining one-fourth would look like the second scan line in that figure.

Finally, Figure 2.10 is a photo of a vectorscope displaying NTSC split field color bars. All luminance information is excluded, leaving only the color subcarrier amplitude and phase information. Note that the

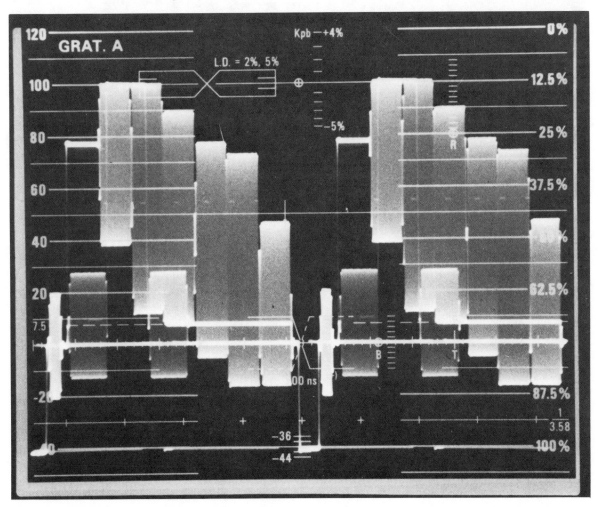

Figure 2.8: Normal waveform monitor presentation of two horizontal scan lines of NTSC split field color bars. Courtesy Tektronix, Inc.

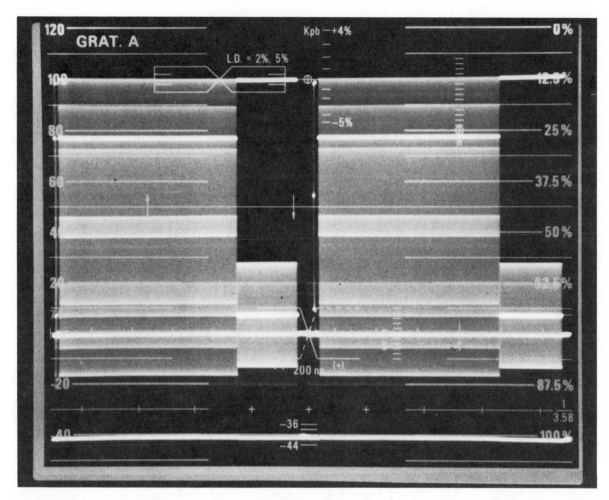

Figure 2.9:   Normal waveform monitor presentation showing two complete fields (one complete frame) of NTSC split field color bars. Courtesy Tektronix, Inc.

color burst has been properly adjusted at 180° (nine o'clock) and at the 75% calibration mark on the vectorscope display. Once these adjustments are made, all primary and complementary colors can be analyzed for the proper phase and amplitude.

The function and the adjustment of these signals is discussed further in Chapters 3 and 6.

The PAL Color Television Signal

Essentially, the 625-line PAL color television signal is a variation of the 525-line NTSC television signal. The phase alternating line (PAL) system was developed some ten years after the NTSC signal, with the idea of retaining all of the best NTSC features while correcting NTSC's greatest drawback: susceptibility to chroma phase (hue) shift.

The vacuum tube circuitry used in early color TV

sets was much less stable than today's solid state circuitry. Consequently, vacuum tube sets operating on the NTSC standard often suffered from a minor malady called subcarrier burst phase drift. On the TV screen, phase drift appeared as color hue changes, particularly in the skin tones and blue-green areas of the picture.

Phase Reversal

During the development of PAL, Dr. Walter Bruch of Germany's Telefunken Research Laboratories decided to experiment with a controlled, line-by-line 180° subcarrier phase shift approach to eliminating hue shift problems. This was a revival of an idea originally put forth by B.D. Loughlin in 1950, during the debate that preceded the setting of NTSC signal standards. Basically, Loughlin's idea called for a complete field phase switching to eliminate hue shift. Unfortunately, it was found that, using the TV technology available in the early 1950s, Loughlin's field switching method introduced severe flicker into the picture.

Dr. Bruch's phase alternating line approach eliminated this problem. The PAL system reverses the subcarrier phase on *each scan line,* nullifying any hue shift problems by comparing the phase of one scan line with the phase of the next. Figure 2.11 is a drawing of a vectorscope screen displaying PAL signal information. As the picture shows, the line-by-line 180° phase shift of the PAL subcarrier frequency creates a mirror image about the U-axis (0° to 180°). Notice that the bracketed boxes with capital letters designating the color coordinates are mirrored by unbracketed boxes designated by lower-case letters representing the same color coordinates of the reverse phased subcarrier. Note, also, that the reference burst is at a 45° angle relative to the normal NTSC setting and swings

a full 90° around that setting due to the subcarrier phase reversal. This results in an average burst phase position of 180°—the normal NTSC setting.

All this adds up to a fairly effective way of eliminating hue shift. As a consequence of the mirror imaging of the color coordinates, a phase shift of even +10° will be countered by a −10° shift on the next scan line. This comparison correction results in no apparent phase shift.

Because PAL uses this automatic color correction system, no hue controls are needed on PAL television receivers. In actual practice, drastic phase shifts appear on the TV screen not as hue changes but as saturation changes. This is due to what is called the vectoral averaging effects of the subcarrier phase alternation, a full explanation of which would require mathematical calculations that are far too involved to cover in this book.

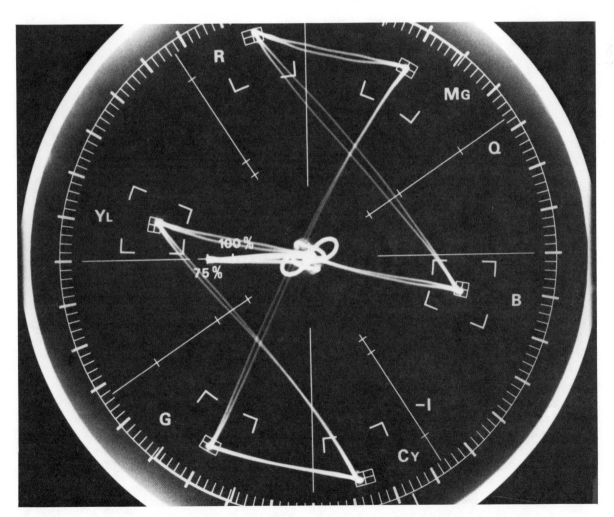

Figure 2.10:   Normal vectorscope display of properly adjusted NTSC split field color bars. Courtesy Tektronix, Inc.

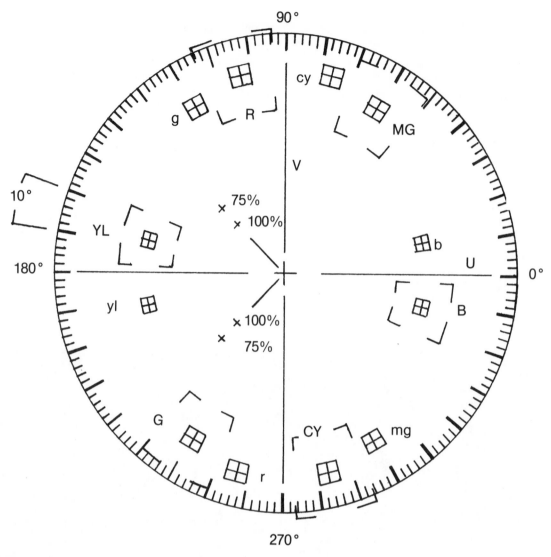

Figure 2.11: Vectorscope displaying PAL signal information.

PAL and NTSC

In most other aspects, the PAL television signal closely resembles the NTSC signal. For example, both PAL and NTSC use the same sort of interlaced scanning system, the only real difference being that the scanning beam in the PAL system scans 312.5 lines per field at a 50 Hz rate. PAL also employs a different scan line numbering method, with the numbering for the 625 scan lines starting at the first sync edge of field one and continuing consecutively through both fields.

Figure 2.12 illustrates two horizontal scan lines of the PAL television signal. Compared to the NTSC waveform display in Figure 2.3, the most noticeable difference is the absence of black pedestal. This is because in PAL the blanking level and reference black

level are one and the same. Figure 2.12 also shows that PAL luminance levels are measured in millivolts (mv) rather than IRE units.

Color saturation and hue information are impressed upon the luminance signal in the same manner as used in the NTSC system, with the exception that PAL uses a 4.43 MHz subcarrier frequency and, of course, phase reversal on each horizontal line.

The PAL horizontal and vertical blanking intervals serve the same purpose as in NTSC, but as Figure 2.13 shows, there are the following differences in the PAL vertical blanking interval compared to the NTSC vertical blanking interval shown in Figure 2.6:

• The length of vertical blanking is typically 26 horizontal lines (compared with 19 to 21 lines in NTSC);

• The length of vertical sync is 2.5 horizontal lines (compared with three lines in NTSC);

• There are five pre-equalizing and postequalizing pulses of half a horizontal line duration preceding and following the vertical sync pulses (compared with six equalizing pulses in NTSC); and

• A special burst blanking known as the "Bruch Blanking System" is used in the vertical blanking interval to assure that the subcarrier burst phase is the same at each end of each blanking interval. The Bruch Blanking System was invented by Dr. Walter Bruch to alleviate the color "jitter" at the top of the picture caused by the phase reversal that occurred through the vertical blanking period.

Use of PAL

The PAL television signal is used in more countries than any other television standard, as indicated in Appendix A of this book. Due to variations in transmitter bandwidth specifications, many countries actually use one of the three modified PAL systems: PAL-I, PAL-B or PAL-G. To make things even more confusing, Argentina and Brazil broadcast their own versions of PAL. Argentina transmits using PAL-N, which is the same 625-line 50 Hz system as the three modified PAL systems, except that the subcarrier frequency is 3.58 MHz. Brazil transmits on a system called PAL-M, which is a 525-line 60 Hz system using a 3.58 MHz subcarrier frequency similar to NTSC, except that it employs PAL subcarrier switching methods.

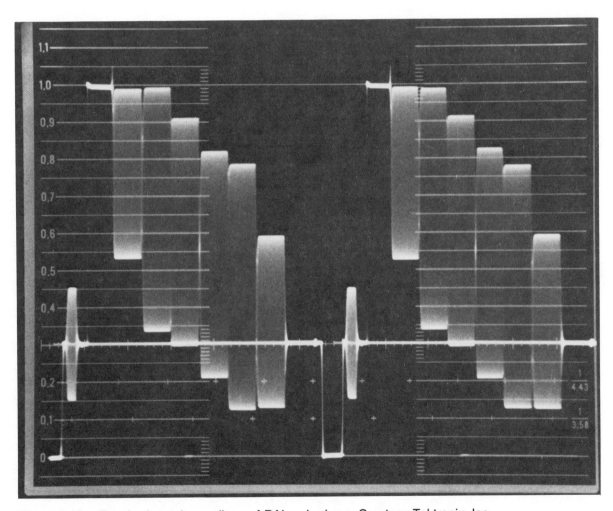

Figure 2.12: Two horizontal scan lines of PAL color bars. Courtesy Tektronix, Inc.

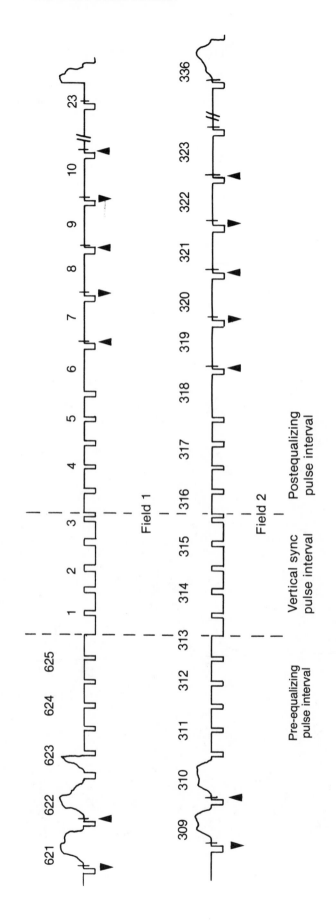

Figure 2.13: The 625-line PAL vertical blanking interval.

THE COMPONENT VIDEO SIGNAL

The component video signal is different from the composite video signal (either PAL or NTSC) in the way the signal is processed. While composite video is processed as a single (one wire) signal containing the combined luminance signal, chrominance signal and sync information, component video is processed as separate (two or three wire) signals containing a separate luminance signal and either one or two separate chrominance signals. From the beginning, component video has been the basis for all color camera design. Color cameras work on the principle that the light value of an image entering into the camera lens is analyzed and electronically transformed into its primary colors of red, green and blue (RGB).

Although some early TV equipment such as RGB type chroma keyers used these separate signals, the more common method was to encode the signals into the familiar composite video form. Component video processing begins by taking the RGB signal and generating a luminance signal (Y) that carries the information about how much light is present in each area of the image. Further processing results in generating two separate chrominance signals, usually called R-Y (R minus Y) and B-Y (B minus Y) that carry information about which color (hue) and how much color (saturation) is present in each area.

The advantage of component video is its durability. Component video maintains a cleaner video signal throughout the various stages of production recording, editing, duplication and distribution. Component key signals and RGB chroma-key signals maintain much sharper edges than composite key signals because they are not subject to signal distortion problems such as busy key edges, chroma crawl, ''chroma rainbowing'' and loss of image detail in colored areas.

Because component-based television equipment

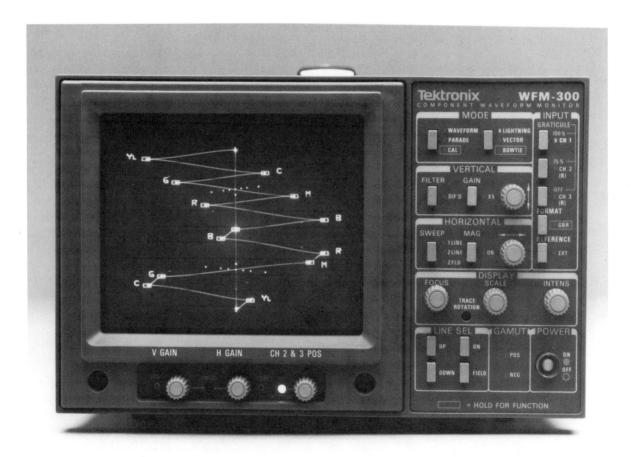

Figure 2.14: NTSC color bars as seen on a component television waveform monitor. Courtesy Tektronix, Inc.

processes the video signal separately, it produces signals that are very different from composite television signals. Figure 2.14 shows the standard NTSC color bars as seen on a component television waveform monitor. This "lightning display" presentation shows the relative timing and amplitude data for all three signals simultaneously. The display is generated by plotting each of the color difference points (B-Y, R-Y) against its luminance component (Y), using the luminance signal as the reference. The upper portion of the display corresponds to the B-Y signal, while the lower portion corresponds to the R-Y signal. In actual use, the timing relationships of the Y to B-Y signal, and the Y to R-Y signal, can be verified by comparing the path between the G and M points for both the upper and lower portions of the display. This represents the edge between the green and magenta bars in the color bar signal. The proper timing relationship results when the G-M line intersects the middle dot of the central path. Any deviation from this intersection represents a relative timing error in the system.

The proper amplitude for each channel is indicated when each color bar point is located inside the appropriate box indicated on the waveform monitor graticule overlay.

Component video equipment may be interconnected in three different ways: the luminance and separate chrominance signals can be arranged in a three wire connection; the luminance and multiplexed chrominance signals (see Betacam® VTR) can be arranged in a two wire connection; or the encoded composite video signal can be used in the familiar one wire connection.

THE TELEVISION ASPECT RATIO

Editors also need to be aware of the television aspect ratio, the standard ratio of picture height to width. In the NTSC 525/60 TV standard, the aspect ratio is 3:4 (see Figure 2.15). That is, for every three inches of height, the scanned area measures four inches across (9″ x 12″, 12″ x 16″, etc.).

However, due to transmission losses and TV receiver misadjustments, most home TV sets do not display the entire scanned picture. To avoid recording information in the cutoff zone around the edges of the TV picture, video engineers developed two related standards: the safe title area and the safe action area (also shown in Figure 2.15). The safe title area corresponds to the film "8 field" (80% of the picture area), and it marks the boundary of the zone safely within the readable scanning area of home receivers. The safe action area corresponds to the film "9 field" (90% of picture information)—the absolute boundary for ensuring that the action displayed on a TV camera or monitor is not cut off by the home TV screen. Both boundary areas are of critical concern when a video editor is working with still or motion graphics.

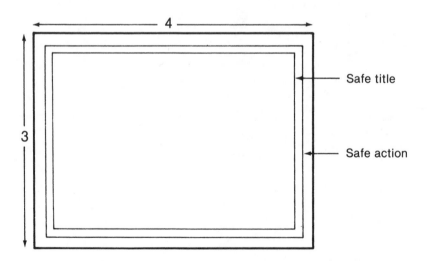

Figure 2.15: The television aspect ratio, safe title and safe action areas.

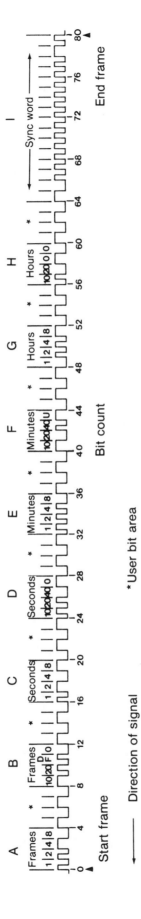

Figure 2.16: One complete line of SMPTE/EBU longitudinal time code.

THE THEORY AND PRACTICE OF SMPTE/EBU TIME CODE

No aspect of videotape recording seems to create as much confusion as the subject of SMPTE/EBU time code (commonly called "SMPTE code"). Despite the tremendous advantage of time-code editing over the older methods of editing videotape discussed in Chapter 1, recording with SMPTE time code still remains one of the most misunderstood processes for those entering (and for some of those already working in) the videotape recording industry. The following pages provide a basic overview of time code—an overview that should help clear up some of this confusion.

SMPTE time code is indexed in hours, minutes, seconds and frames (30 frames per second in NTSC and 25 frames per second in PAL). NTSC time code readouts range from 00:00:00:00 to 23:59:59:29, recycling at each 24-hour interval. There are two forms of SMPTE time code. The most common form, known as longitudinal (or "serial") time code (LTC), is recorded along an audio track on the videotape. The second form, known as vertical interval time code (VITC), is recorded in the vertical interval during the production shoot. In both cases, SMPTE code is recorded as a binary signal designed to generate a complete line of information twice during each television frame (one complete line of information per field).

Basic Binary Code

Although a detailed analysis of binary coding is beyond the scope of this book, a basic introduction should help video editors understand how time code is generated. Figure 2.16 illustrates one complete line (or frame) of SMPTE longitudinal time code. This sample frame will serve as a model in our discussions of binary code, as well as in our subsequent discussions of SMPTE time code.

Basically, the binary code is an electronic signal that switches from one voltage level to another, forming a chain of voltage pulses. Each pulse is called a bit. As Figure 2.17 shows, a bit is designated as having a value of "0" (zero value) if it is a complete pulse or a value of "1" (positive value) if the voltage level changes at mid-pulse (forming a half pulse). The time code line for one video frame contains a total of 80 individual bits.

By organizing the bit pulses into groups of four (called "four-bit words") and assigning representative values to each pulse, we are able to generate a string of pulses that carries specific time code information. For example, Figure 2.18 shows a slice taken from the complete time code line containing the hours, minutes, seconds and frame information for one video frame illustrated in Figure 2.16. The slice shown includes the information slots ("words") that will deter-

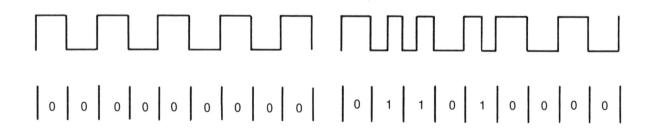

Simple binary code containing all complete pulses, yielding values of all zeros.

Simple binary code containing complete pulses and half pulses, yielding values of zeros and ones respectively.

Figure 2.17: How binary code is represented in SMPTE longitudinal time code.

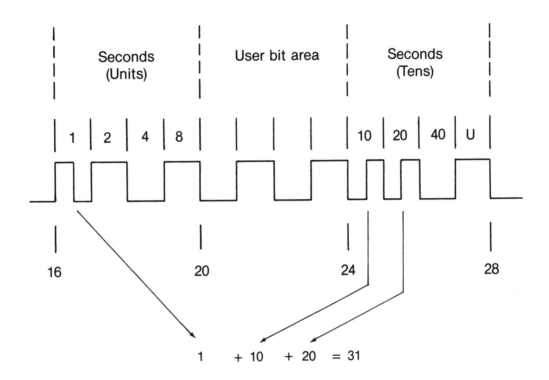

Figure 2.18: Expanded view of bits 16 through 28, representing the "seconds" and "tens of seconds" slots of SMPTE time code.

mine the number of seconds, in both units and tens, that will appear on the time code display. Bits 16 through 19 are the four-bit word that forms the "units" slot, and bits 24 through 27 are the four-bit word that forms the "tens of seconds" slot. By changing the voltage values of individual bits from pulses (a value of zero) to half pulses (a positive value of 1), we can change the seconds and tens of seconds count on the time code read out. *Only the values of the slots containing the half pulses are counted when we add up the bit totals.*

Figure 2.18 also suggests how this system of assigning and adding up voltage values works. As the illustration shows, the units slot (bits 16 through 19) contains only one half pulse, which fills the slot carrying a value of one unit. However, in the tens of seconds slot, both the 10 second and 20 second slots are at

half-pulse positive value. By taking the 1-second, 10-second and 20-second bits (the three slots that are at half-pulse positive value) and adding them together, we get:

$$1 + 10 + 20 = 31 \text{ seconds.}$$

In other words, the "seconds" on our time code display would read "31."

That, basically, is how binary time coding works. Of course, in actual practice, you will never be called upon to decipher the bits and words that form SMPTE time code since the code is generated, read and displayed electronically.

Components of Time Code

As Figure 2.16 shows, and as the preceding discussion suggests, the 80-bit longitudinal time code is

---

Table 2.1: Computing Time Code from Time Identification Slots

---

Tens of hours (10, in slot H) plus hours (2 + 4 in slot G) equals 16 hours
Tens of minutes (40, in slot F) plus minutes (1 + 2 + 4, in slot E) equals 47 minutes
Tens of seconds (10 + 20, in slot D) plus seconds (1, in slot C) equals 31 seconds
Tens of frames (20, in slot B) plus frames (1 + 2, in slot A) equals 23 frames

---

composed of a series of four-bit words (or "slots"), each of which carries a different piece of information:

A. Frame number (units)
B. Frame number (tens)
C. Seconds (units)
D. Tens of seconds
E. Minutes (units)
F. Tens of minutes
G. Hours (units)
H. Tens of hours
I. Sync word (containing synchronizing cues)

In addition, the 80-bit time code shown in Figure 2.16 contains eight "user bit areas," indicated by asterisks. All of these information slots—plus one other, the dropped-frame bit—are described in more detail below.

Time Identification Slots

The eight four-bit words (slots A through H) that carry information about hours, minutes, seconds and frames are called time identification slots. Table 2.1 shows how we can "read" the information in the slots in Figure 2.16 by adding the half-pulse values. Displayed on a monitor screen, our sample time code would actually look like this: 16:47:31:23.

User Bit Slot

SMPTE code provides eight optional four-bit words, the user bits, that can be used for inserting any information deemed necessary during production. As Figure 2.16 shows, these four-bit words are placed between the time identification slots. They can accommodate enough digital data for four alphabetical characters, eight numerical characters or a combination of the two.

Producers might use the slots to insert information about one or more of the following: reel numbers, take numbers, editing dates, playback code from audio tape machines, or field-rate time code for film-to-tape transfers.

Sync Word Slot

Bits 64 through 79 contain a constant pattern of "zero" and "one" bits known as the sync word. Its job is to indicate whether the tape is traveling in forward or reverse and to identify the end of each frame and the beginning of the next. Computer editing systems look at *both* the time code and the video signal during the editing process. To ensure proper coordination between the two, editors must make sure that the time code generator is synchronized with the video signal as the time code is recorded onto the videotape. If the sync word in the time code tells the computer that the video frame starts in one place and the video signal indicates it starts in a different place, the computer system will simply refuse to perform the edit.

Dropped-Frame Bit

Bit 10 of the time code is designated as the dropped-frame indicator. When set to "1," this bit indicates to all code-reading devices that the time code is deleting those numbers necessary to maintain "real-time accuracy"—called drop-frame code. When it is set to "0," bit 10 indicates that all numbers are present in the code—called non-drop-frame code. Drop-frame and non-drop-frame codes are explained in the next section.

NON-DROP-FRAME AND DROP-FRAME TIME CODE

There are two operating modes for 30-frame-per-second SMPTE time code: non-drop-frame and drop-frame. Non-drop-frame time code (also referred to as "color time code") is a continuously counting code that indicates hours, minutes, seconds and frames and that resets at each 24-hour interval. Drop-frame time code (also referred to as clock-time code) has one significant difference. It drops, or skips, frames at designated time intervals. It does this so the time code will maintain real-time accuracy.

As you will recall from our description of the TV signal earlier in this chapter, the NTSC standard television signal operates at 59.94 Hz, not the 60 Hz at

Figure 2.19: EECO vertical interval time code generator/reader model VIG-850. Courtesy EECO, Inc.

which clocks operate. As a result, the clock and the TV signals are actually running at slightly different speeds. This leads to the sort of problem you encounter when you start two 1/4-inch audio recorders in sync but run them at slightly different speeds. After a short time, they begin running out of sync, at a gradually expanding rate.

The same is true with SMPTE time code and the TV signal. The two may start in sync with the clock, but, because of the speed difference just described, there will be an expanding elapsed time difference between the clock and the time code readings. The actual difference amounts to 108 frames per hour, or 86 seconds for each 24-hour period. To correct for this time difference, drop-frame time code is designed to skip two frames a minute except for every tenth minute.

For a number of reasons, editors using time code with noncomputerized editing systems are probably better off using non-drop-frame code. While the non-drop-frame code is not real-time accurate, it is considerably easier to calculate manually than drop-frame code. Just keep in mind that at the end of a 30-minute project, the elapsed time of the program will be one second plus 24 frames longer than the length generated by the time code display.

An industry rule of thumb is that, for each half hour of programming, a program using non-drop-frame time code is actually two seconds longer than the time code indicates. Therefore, if the program is designed to be 28 minutes and 39 seconds long, the calculated non-drop-frame time should only equal 28 minutes and 37 seconds.

USING VERTICAL INTERVAL TIME CODE

As mentioned earlier, there are two types of time code: longitudinal time code (LTC) and vertical interval time code (VITC). VITC was developed after LTC to correct two of the longitudinal code's limitations—its tendency to misread the time code when the tape is moving at very slow speeds and its inability to read the code at all in still-frame. However, VITC also has its faults, particularly its habit of becoming unreadable at fast-motion speeds. As a result, when using VITC, video editors usually have both VITC and LTC recorded on their workprints.

Adding VITC and LTC to Tapes

VITC generator/readers, such as the EECO VIG-850 shown in Figure 2.19, insert the time code into any line

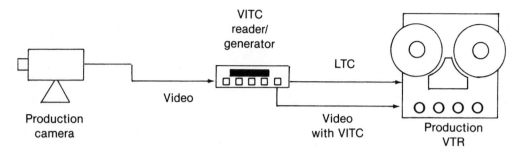

Simplified method of adding vertical interval time code (VITC) and longitudinal time code (LTC) during production.

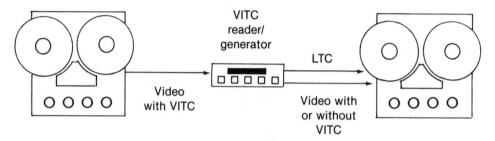

Simplified method of adding matching LTC while duplicating a videotape previously coded with VITC.

Figure 2.20: Two ways of recording time code.

between lines 10 and 20 of the vertical blanking signal. If desired, the VITC generator/reader can also insert LTC into the appropriate VTR audio channel.

Figure 2.20 demonstrates, in simplified fashion, two ways of recording time code with the VITC generator/reader. The first example illustrates adding both VITC and LTC to the tape on the production shoot by feeding the video from the production camera into the generator/reader. The generator/reader then adds the SMPTE time code signal to the video signal and sends the video signal with both VITC and matching LTC to the production VTR. Keep in mind, however, that only one video signal may be processed through the VITC generator/ reader at any one time, making it necessary to use one generator/reader for each individual production video feed.

The second part of Figure 2.20 shows the procedure of making a dupe (duplicate) of a VITC-coded videotape and then adding matching LTC to the dupe's audio track, using the reader/generator feature of the time code equipment.

Vertical interval time code is normally intended for use from still-frame to approximately 45 times normal speed, that is, as long as the VTR maintains a usable picture in the shuttle mode. For speeds above the

VTR's usable picture range, the time code reader switches from VITC to LTC.

Components of VITC

Vertical interval time code is very similar in composition to longitudinal time code. There are, however, extra bits known as "housekeeping" bits that bring the VITC "line" to 90 bits per frame (as compared to 80 bits per frame with longitudinal time code). These extra bits are described briefly below.

Sync Bits

Two sync bits precede each of the nine data bit groups. These are always set to "1" and "0" values, respectively.

Field Bit

Bit 35 is known as the field bit. When this bit is at "0" value, it denotes the first television field. At "1" value, it denotes television field two. Remember, SMPTE code numbers are generated twice each frame, or once each field.

Cyclic Redundancy Check Code (CRC)

In VITC, bits 82 through 89 store an error detection code in place of the sync word. The pattern of these bits is compared to a known standard in the VITC reader to insure that all bit data have been properly generated.

Advantages of VITC

In addition to its ability to provide a readable display in freeze-frame and at very slow tape speeds, VITC offers the following advantages over LTC:

• It requires no special amplification or signal-processing equipment during playback or duplication;

• It provides field-accurate access (as opposed to the frame-accurate access available through LTC);

• It does not occupy an audio track, allowing for multitrack audio recording if desired; and

• An error detection code (CRC) provides protection against reading errors.

Remember, however, that it is impossible to precode VITC on tapes that will be used as editing master stock, because the vertical interval code must be inserted into the actual program video signal.

FIELD-RATE SMPTE TIME CODE

Field-rate SMPTE time code was developed by Gray Engineering Laboratories, Inc., as a means of editing film workprints on video and then conforming the negative on film. It is designed to help overcome the frames-per-second difference between film (24 frames per second) and NTSC video (30 frames per second).

THE SEVEN MOST COMMON QUESTIONS ABOUT SMPTE TIME CODE

When should time code be recorded?
Whenever it is feasible, SMPTE time code should be recorded during the production session. I say this for three reasons:
1) The cost per day of leasing a SMPTE code generator/reader unit during production is less than half the cost *per hour* of having each reel of videotape time coded ("striped") after production.
2) In multiple-camera productions where the video signals from individual cameras are recorded separately, each tape should have identical time code for editing efficiency. Although it is possible to edit tapes with different time code numbers or to "sync up" reels in post-production and record identical time code on each, these alternatives add extra time and expense to the editing process.
3) Recording the time code during production allows the producer to obtain an immediate, accurate logging of selected scenes, saving much of the time required to screen reels that are time coded in post-production.

If it is necessary to time code tapes after the production session, will I have to record down a generation?
No. You simply record the time code on the appropriate audio track of the original master reels.

What happens if the code recorded during production is not good?
Two basic problems can occur when time code is recorded during production:

• The code level is recorded too low;
• The code generator is not synchronized with the video signal, causing code drift.

Both problems have a relatively simple solution. In both cases, you can try "slaving" a code generator and restriping the reels. Slaving is the process of feeding the bad code signal to a code generator with slave capability (any post-production facility should have this capability). Once the code generator is put in slave mode and the VTR is put in play mode, the generator will put out fresh code with the identical numbers being fed to its slave input. Then the videotape operator takes the generator out of slave mode and pushes the proper record button on the VTR. With the VTR in the record mode, fresh time code is placed on the original master reel.

A word of warning: Make sure that this slaving process is repeated each time the video signal on the original reels breaks up (indicating tape stops and starts during production).

(continued on next page)

THE SEVEN MOST COMMON QUESTIONS ASKED ABOUT SMPTE TIME CODE (continued)

Is it all right to record time code from one tape onto another without using a slave generator?

Unless you are using vertical interval time code, it is not advisable to duplicate time code "VTR to VTR" without either slaving a time code generator or using a code regeneration amplifier. SMPTE time code is a series of square wave pulses that become ragged when magnetically reproduced in the playback process. When duplicated over a few generations, this distortion of the square wave signal can cause misreading.

Most post-production facilities either use slave code generators or have time code regeneration amplifiers installed in their VTRs. Time code regeneration amplifiers increase the level of the ragged time code signal and then clip the top edge at the proper voltage level, restoring the sharp, square wave edge of the signal.

Should I use drop-frame or non-drop-frame code?

This really depends on the type of editing system that you'll be using. As I stated earlier, I prefer drop-frame time code, since the elapsed time is accurate clock time. However, on editing systems that do not calculate drop frame automatically, I prefer non-drop-frame code.

At what numbers should time code be started?

Most productions are recorded and edited with the code corresponding to the time of day. However, some producers prefer the SMPTE hour code to equal the reel number. This generally works well on film-to-tape transfer reels and on extended, continuous tapings that run under 60 minutes. One drawback to this method is the fact that SMPTE code only goes up to 24 hours. As a result, the number of reels that can be numbered in this way is apparently limited to 24. However, you can get around this problem by assigning reel numbers in the hundreds on the second 24 reels (101-124, 201-224, etc.).

Technically it does not matter where the code numbers start. But from a convenience standpoint, beginning the project on the edit master reel from the start of an hour or minute makes overall timing of the project considerably easier. I have found that some editors prefer to start their code at one hour (01:00:00:00), and that other editors prefer ten hours (10:00:00:00). Either way, it's strictly a matter of preference and convenience.

Why not start at zero? Actually, you *can* start the time on the edit master reel at zero (00:00:00:00), provided the following facts are taken into consideration.

• The first edit should be made at 23:59:59:29 (one frame before 00:00:00:00). This will prevent the problem that pops up when the VTR rewinds to preroll for the first edit. If the first edit point is set at "0," prerolling will cause the tape to cross into the 23 hours time code. When this happens, the VTR will start rewinding the tape toward zero.

• If the first edit is very short, you might encounter this preroll problem on the second edit also. Just be sure you have enough preroll room to avoid crossing over into the 23 hour area.

Why should I use time code?

To review, the advantages of using SMPTE time code, whether longitudinal, vertical interval or field-rate, are listed below.

Accuracy and repeatability. Time code permits edits that are accurate down to one frame or field and that are repeatable. Edit points can also be adjusted in one frame increments.

Precise time reference. With time code, editors can calculate scene duration quickly and accurately, simply by subtracting the "edit-in" point from the "edit-out" point. Time code also gives editors frame-accurate running times throughout the editing process.

Videotape recorder/audio tape recorder synchronization. Time code allows editors to synchronize videotape recorders with audio tape recorders, permitting multiple audio track dubbing.

Adding special effects in the editing process. Using SMPTE time code, editors can add intricate digital effect sequences, operate video switcher functions, roll remote audio recorders or activate any other audio or video sources that can be triggered automatically by a time code comparator device.

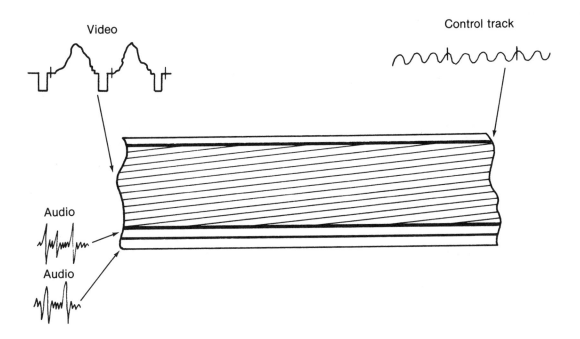

Figure 2.21: Simplified diagram of a helical videotape recording.

Field-rate code is very similar to normal frame-rate SMPTE time code, with two exceptions: it is generated at a field rate rather than a frame rate, and it is designed to permit film-count information to be recorded in the user-bit area. Further information on field-rate time code is available in several of the books and articles on video-assisted film editing listed in the bibliography.

BASIC VIDEO RECORDING

Now that you've received a basic introduction to the video signal and SMPTE time code, you might be wondering how all that information is recorded on the videotape.

Getting It on Tape

A videotape recording consists of a video portion, a control track with frame pulses, and one or more audio channels (see the simplified diagram of a helical recording in Figure 2.21). The video portion of the recording consists of picture information combined with horizontal and vertical synchronization pulses (the composite video signal, discussed earlier in this chapter) impressed on a very high frequency signal known as a carrier frequency.

The control track consists of a continuous, stable, low-frequency signal. When the tape is played back, the control track signal stabilizes the VTR's servo system. In a way, it is analogous to sprocket holes on film. Combined with this control track signal are a series of sharp pulses, called frame pulses, that occur at the beginning of each frame at the same time as the vertical synchronization pulse for field one of that frame. (Thus, while vertical sync pulses occur 60 times per second, frame pulses occur 30 times per second.)

On each video recording, there may be one, two or three audio channels, depending upon which format videotape machine is being used (see the section on videotape recorder formats later in this chapter). Time code is usually recorded on one of the audio tracks. Typically, the time code is recorded on the audio cue track with 2-inch (quad) machines, on audio track 3 with 1-inch machines, and on audio track 2 with cassette VTRs. Time code may also be recorded on a special "address track" available on BVU-series cassette machines, thus leaving both audio tracks free for multiple-track recording.

To make sure they are placing "clean" signals on the tape, editors should check to see that the VTRs are properly aligned, that they are generating optimal recording currents, and that the tape transport paths are free of dust and dirt. Failure to maintain these standards will result in a loss of recording quality.

Videotape Basics

Videotape consists of a very fine layer of oxide coating on a mylar-based film backing. Since the entire recording surface is only 0.0002 of an inch thick, obstructions as small as a particle of cigarette smoke can cause problems. It is therefore imperative that precautions be taken to keep the videotape stock clean. Smoking should be prohibited in areas where tape is used or stored. Tape should be handled by touching only the protective mylar backing, *not* the oxide coating. Tape should be stored in a cool, dry, dust-free place, away from sources of magnetic fields, such as electric motors, fluorescent light fixtures and electric transformers.

Also, when using helical VTRs that keep the tape next to the video heads at all times, editors should be careful that the VTR does not remain in freeze-frame mode for more than a few minutes. When tape is in the freeze-frame mode, the video heads keep scanning the same section of tape, causing friction heat and excess wear that can result in either video drop-outs or an unstable video signal. To guard against this problem, many helical VTRs are designed with automatic timers that shut off the video head motor after about 40 seconds of operation in the freeze-frame mode.

Finally, it should go without saying that video editors should not use bent and warped reels, since the edges of the reel will nick the sides of the videotape and cause edge damage (a recurring problem in video post-production).

VIDEOTAPE RECORDER FORMATS

Currently, the video universe is populated by a confusing variety of videotape recorder formats. Since a video editor often has no way of knowing in advance which format he will be working with on a particular project, it makes sense for editors to become familiar with as many different VTR formats as possible. Videotape recorder systems used in professional video editing include the 2-inch (quad) format, the 1-inch type B format, the 1-inch type C format and the 3/4-inch format. Recently, Beta and VHS format machines (1/2-inch cassette machines) have also begun showing up in editing bays, particularly in offline workprint applications.

VTR formats are determined by several distinguishing characteristics:

• The width of the videotape used with the VTR (2-inch, 1-inch, 3/4-inch, 1/2-inch);

• The path the videotape takes as it makes its way past the video heads and through the machine (quad-

Figure 2.22: Ampex AVR-3 2-inch broadcast videotape recorder. Courtesy Ampex Corp.

ruplex, helical scan, type B format, type C format, Beta, VHS);

• Whether the VTR features segmented or unsegmented video; and

• The speed of the tape as it passes through the VTR.

Operating instructions for different VTRs can be found in the manuals that accompany individual makes and models. In this section, I'll discuss the differences among the various formats as well as some of the factors that determine which format is appropriate for a given editing application.

Quadruplex (Quad) VTRs

Quadruplex VTRs employ magnetic video recording tape that is two inches wide and that moves through the machine at a recording or playback speed of 15 inches per second. (Recording at a speed of 7.5 inches per second is possible with the use of a different video head, but this speed is rarely in use today.) Figure 2.22 shows one popular quad VTR, the Ampex AVR-3. A closer view of the AVR-3 tape transport, illustrating the path the videotape takes as it moves from the supply reel to the take-up reel, is shown in Figure 2.23. The videotape is threaded so that, from the supply reel, it moves past the tension arm assembly and across the video erase head assembly, past the video head assembly and audio head assembly, around the capstan, past the take-up side tension arm assembly, and onto the take-up reel.

The vacuum tape guide on the video head assembly holds the videotape steady as it passes through the video head assembly. The capstan is extremely stable, and the vacuum mechanism eliminates the need for the capstan pinch roller used in earlier model VTRs.

The quad video head assembly contains four video heads mounted 90 degrees apart on a rotating head

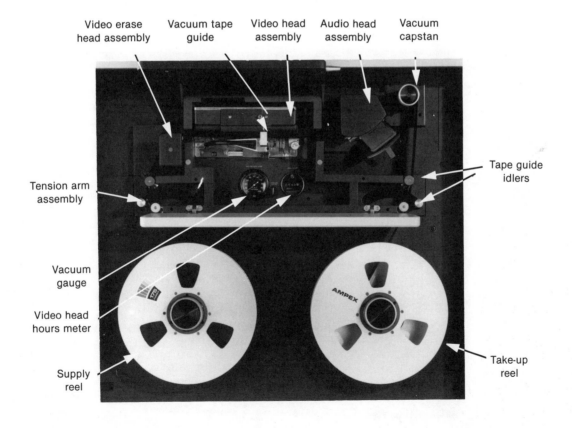

Figure 2.23: Tape transport path for the Ampex AVR-3. Photo courtesy Ampex Corp.

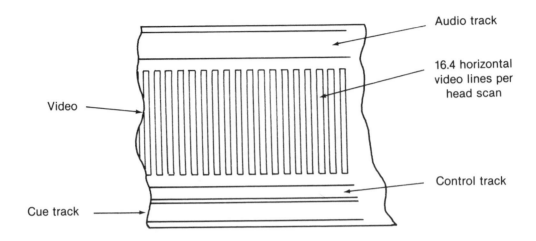

Audio track

16.4 horizontal
video lines per
head scan

Video

Control track

Cue track

Figure 2.24: Video and audio tracks on a developed section of 2-inch videotape.

wheel. Each head passes across the videotape, recording (or reproducing) 16.4 horizontal lines of video information as the tape is pulled past the head wheel.

As the following equations show, the quad VTR scans one complete field of video information for every four complete revolutions of the video head wheel.

Figure 2.25:   Bosch Fernseh BCN-51 type B 1-inch videotape recorder. Photo by Darrell R. Anderson.

• Given:

262.5 horizontal lines = 1 video field
16.4 horizontal lines = 1 video head pass
4 video head passes = 1 complete video head assembly revolution

• Therefore:

262.5 horizontal lines ÷ 16.4 horizontal lines
= 16 video head passes
16 video head passes ÷ 4 video heads
= 4 complete revolutions

Figure 2.24 shows the video and audio tracks recorded on a 2-inch tape by the quad machine. Note that the video tracks are at a near 90 degree angle to the direction of tape travel. This is referred to as a "transverse scan recording." Most quad VTRs feature one high-quality audio track and one low-quality cue track, both of which are recorded longitudinally along the tape. The AVR-3 VTR pictured in Figure 2.22 has optional dual channel audio with the addition of a special dual track audio modification kit. The control track with frame pulses is recorded longitudinally along the bottom edge of the videotape, parallel to the cue track.

Since the entire field of video is not scanned by one complete pass of the video heads, the quad format is defined as a "segmented video format." With segmented format machines, slow-motion and still-frame effects are impossible without the use of external electronic devices.

Type B Helical VTRs

The SMPTE standardized type B VTRs employ videotape that is one inch wide and that operates at a recording or playback speed of 9.65 inches per second (245 mm per second). The type B machines manufactured by Fernseh, Inc. (see Figure 2.25) wind the tape onto the take-up reel in what is called a "B wind" configuration. That is, the oxide side of the videotape, which must come in contact with the video head assembly, faces outward on the supply and take-up reels.

Type B machines also use a relatively small video head assembly (called a scanner assembly), which features two rotating video record/reproduce heads and which is designed to operate with the tape wrapped around 190 degrees of the assembly surface (see Figure 2.26). Each video head pass records (or reproduces) 52 lines of video information as the tape moves around

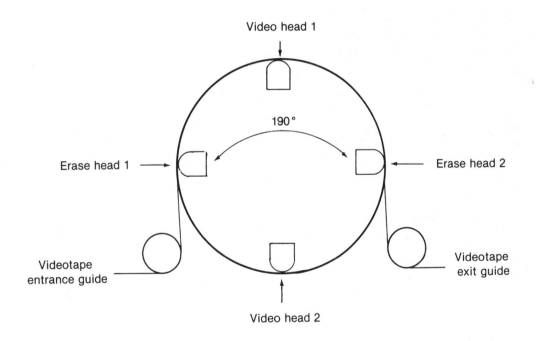

Figure 2.26: Type B video head assembly.

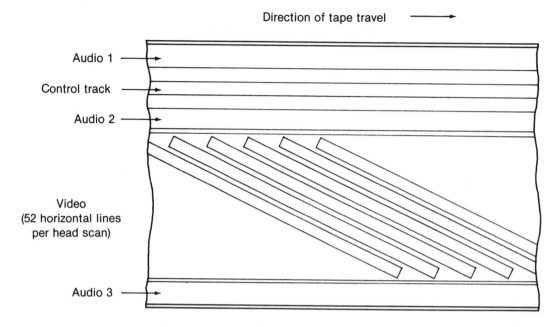

Figure 2.27: Developed section of type B videotape.

Ampex type C 1-inch VPR-2B videotape recorder. Courtesy Ampex Corp.

Sony type C 1-inch BVH 2000 videotape recorder. Courtesy Sony Communications Products Co.

Figure 2.28: Type C helical videotape recorders.

the scanner assembly. The scanner assembly also has two flying erase heads installed for insert editing purposes.

Advantages of the smaller scanner assembly include a reduction of the tension and stress put on the videotape during operation and the incorporation of the entire assembly (including the entrance and exit tape guides) into one unit, which reduces or eliminates interchangeability problems when going from one type B machine to another.

Like quad machines, type B VTRs cannot scan an entire television field with one head rotation. As a result, type B machines are also defined as segmented VTRs, and they cannot offer still-frame or slow-motion capabilities without the use of additional electronic devices. However, the BCN-51 type B VTR shown in Figure 2.25 now features an optional built-in frame store device, allowing the video to be viewed in the jogging and still-frame modes.

Figure 2.27 depicts a developed section of a type B recording. Notice that there are three audio channels and a control track recorded longitudinally on the videotape. Audio channels 1 and 2 are high-quality channels used for recording program audio. Audio channel 3 is a high-quality channel used for program audio, cues and time code recording. Note also the position of the control track. It is placed between audio channels 1 and 2 to allow a relatively high cross talk separation of the audio channels and to protect the control track signal from harm due to videotape edge damage.

Type C Helical VTRs

The SMPTE 1-inch type C helical format was originally devised as a compromise between the early Ampex and Sony 1-inch helical formats, in the interest of compatibility and industry standardization. Figure 2.28 shows both an Ampex VPR-2B type C VTR and a Sony BVH 2000 type C VTR. Although the mechanical designs of the tape threading paths are considerably different, the videotape wrap around the video head assembly of each recorder is identical. Running at a tape transport speed of 9.606 inches per second (244 mm per second), the type C VTR uses a full wrap of videotape around the video head assembly (also referred to as the scanner assembly) and two video record/playback heads inside the assembly. As Figure 2.29 shows, one head is used to record and play back the visible part of the picture. The other video head, called the sync head, is used only for recording and playing back the vertical interval portion of the video signal. For insert editing purposes, there are also two flying erase heads mounted inside the

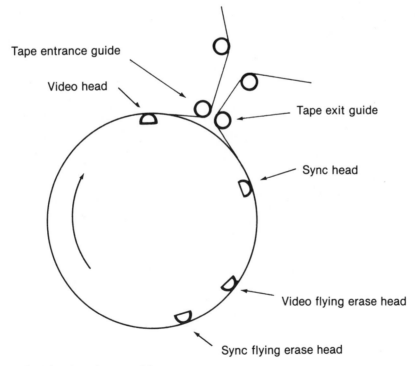

Figure 2.29: Type C video head assembly.

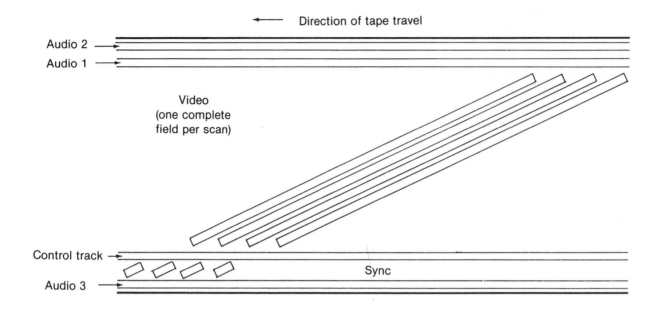

Figure 2.30: Developed section of type C videotape.

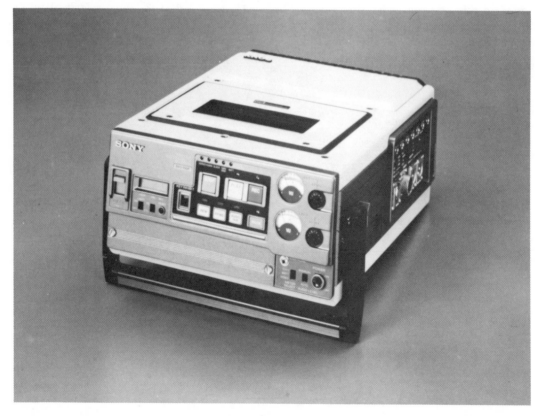

Figure 2.31: Sony BVU-150 portable U-matic® SP videocassette recorder. Courtesy Sony Communications Products Co.

scanner assembly, for erasing the visible picture and the vertical interval portions of the signal.

There are several advantages to the type C format. First, the visible portion of the video signal is recorded by only one head, eliminating visual errors such as banding and skewing that are common in quadruplex recordings. Also, since one revolution of the scanner assembly records or reproduces a complete video field, type C functions as a nonsegmented machine. Unlike quad and type B VTRs, then, type C machines can be operated in slow motion and still frame without the use of external electronic devices.

Figure 2.30 illustrates a developed section of type C 1-inch videotape. Notice that, like the type B machines, type C VTRs feature three audio tracks. Channels 1 and 2 are high-quality tracks used for recording program audio, and channel 3 is a high-quality audio track used for program audio, cues, time code, etc. Audio channel 3 also contains a switchable, high bandwidth pre-amp that provides code readability at fast shuttle rates. Note that the control track has been recorded between the picture video and the sync, so it is protected from physical edge damage to the videotape.

U-Type Cassette VTRs

Sony Corp. introduced the first 3/4-inch U-Matic® recorder in 1971 and subsequently received an Emmy Award for its development. Since then, many offline edit systems, online edit systems and electronic news gathering (ENG) systems have used the 3/4-inch U-Matic videocassette, rather than 1-inch open reel machines, as both playback and edit VTRs. In cassette VTRs, the videotape is enclosed in a plastic box, or cassette, and, upon loading, is threaded automatically. For many years, portable cassette VTRs such as the Sony BVU-150, shown in Figure 2.31, have become increasingly popular in location shooting and electronic news gathering applications. Other high-quality cassette VTRs, such as the Sony BVU-870 and the JVC CR-85OU (see Figures 2.32 and 2.33) have become popular for both offline editing and online cassette mastering applications.

Figure 2.34 illustrates the tape threading path used by most U-type cassette VTRs (except the BVU series, which keeps the videotape fully wrapped except in the eject mode). As the name suggests, U-type video recorders thread the tape in a "U" pattern.

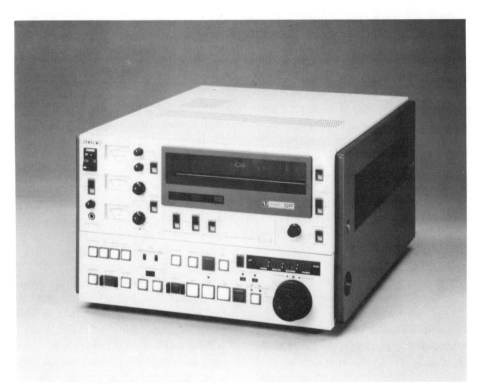

Figure 2.32: Sony BVU-870 SP 3/4-inch videocassette editing recorder. Courtesy Sony Communications Products Co.

Figure 2.33: JVC CR-85OU 3/4-inch videocassette editing recorder. Courtesy JVC Co. of America.

Figures 2.35 and 2.36 show the difference between the regular 3/4-inch cassette recording and the BVU-type recording. Both can be played back on either machine. The only difference is that an optional audio track, called an address track, is recorded on the BVU tape, allowing a longitudinal time code to be placed on the tape across the video recording. This permits the recording of dual track audio when the BVU machine is used with time-code edit systems. Because it is recorded along the unused area of vertical blanking, the time code track does not interfere with the video recording.

Three-quarter-inch videocassette recordings play at a tape transport speed of 3.75 inches per second (95.3 mm per second), and the video head mechanism scans a complete field of video with each head pass. As a result, U-type cassette VTRs record in a nonsegmented video format, and they are capable of slow-motion and still-frame effects. For editing applications, it is advisable to use videocassette stock with still-frame capability. This special tape stock has a heavier gauge nylon backing, so it can better withstand the physical stress involved in the editing process.

Betacam® VTRS

Introduced in 1981, Sony Corp.'s Betacam® 1/2-inch format was a response to the growing demand for even smaller and lighter high-quality ENG/EFP equipment. Betacam® was unique in that its broadcast-quality images were recorded on 1/2-inch videocassette, using either a portable VCR or a camcorder unit which contained the camera and recorder in one light-weight package.

The use of Betacam® studio production VTRs like the BVW-40 (shown in Figure 2.37) has grown in popularity since many major editing facilities now offer multi-format video editing suites. One reason the Betacam® format achieves its high-quality recording is its writing speed, which is almost triple the speed of Sony's consumer Beta format. The higher writing speed gives the unit a much greater recording bandwidth and, along with Betacam's advanced circuit technology, results in greater image resolution and color definition.

Figure 2.38 shows the Betacam® tape pattern displaying the alternating chroma (C) and luminance (Y)

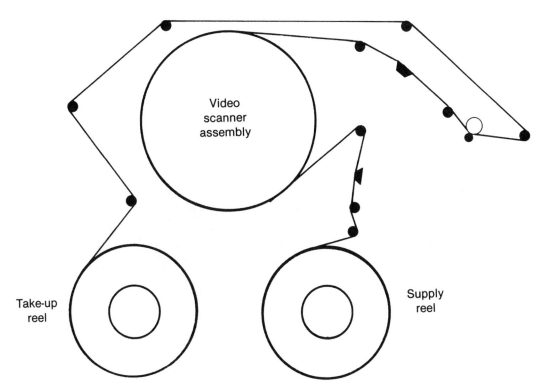

Figure 2.34: U-matic cassette tape transport path in "play," "record" or "pause" modes.

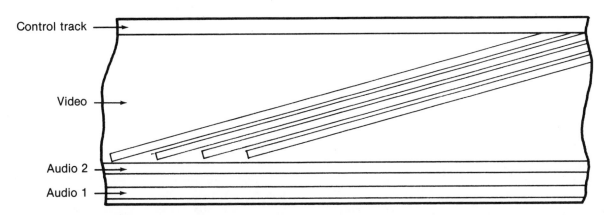

Figure 2.35: Developed section of U-type 3/4-inch cassette tape.

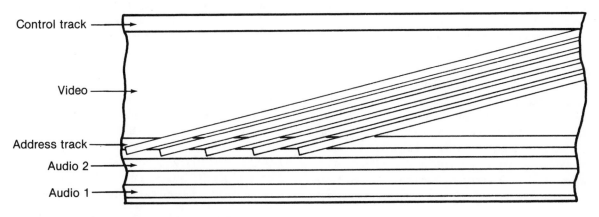

Figure 2.36: Developed section of BVU cassette recording with address track.

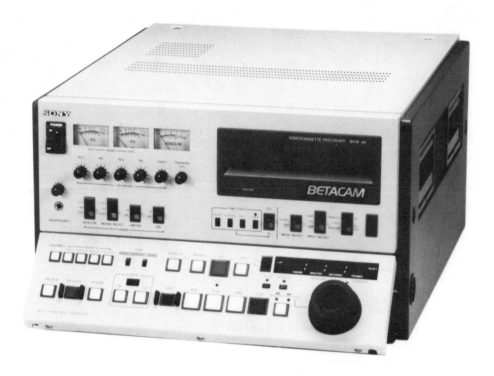

Figure 2.37: Sony BVW-40 Betacam® recorder/editor. Courtesy Sony Communications Products Co.

recording tracks. The two longitudinal audio tracks are positioned along the top of the tape, while the control track and longitudinal time code track are positioned along the bottom of the tape.

The rotating video head assembly, which contains two video record/play heads for the chrominance signal and two for the lumimance signal (see Figure 2.39), records a full field of video information per complete scan. Betacam® processes and records video in the component format and provides outputs in both composite video and component video.

Sony recently introduced Betacam SP® (Superior Performance) production VTRs which are compatible with the current Betacam® format and provide even wider video bandwidths—up to 90 minutes record/playback time and four channels of high quality audio. In addition to the two longitudinal amplitude modulated (AM) audio channels, Sony has added two frequency modulated (FM) audio channels which are

simultaneously recorded with the video information in the rotary video head assembly.

Betacam SP® field production units, such as the portable BVW-35 shown in Figure 2.40 or the Camcorder BVW-505 shown in Figure 2.41, can accept either the regular 30-minute standard oxide Beta tape or the new 30-minute metal-particle tape. The Betacam SP® studio model VTR can accept either the 30-minute metal-particle tape or a 90-minute metal-particle tape.

VHS Hi-Fi VTRs

Due to its tremendous popularity with consumers, the VHS 1/2-inch cassette format is the most popular recording format for professional mass duplication. VHS professional grade VTRs such as the JVC BR-7700U and the Panasonic AG 6500, shown in Figures 2.42 and 2.43, are commonly used in professional mass duplication facilities as well as small manufactur-

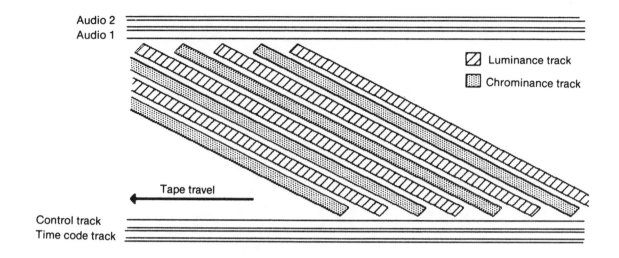

Figure 2.38: Developed section of 1/2-inch Betacam® videocassette tape. Courtesy Sony Communications Products Co.

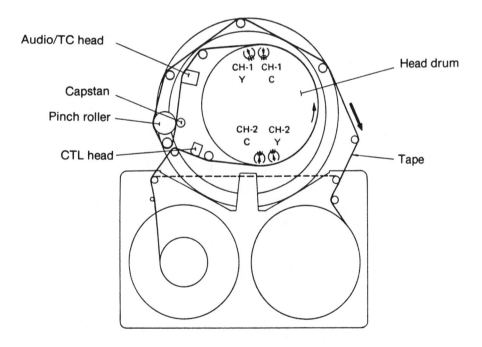

Figure 2.39: Betacam® tape transport path and video head assembly. Courtesy Sony Communications Products Co.

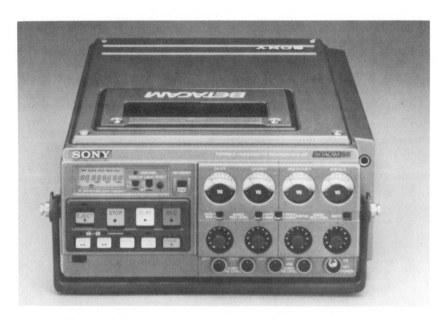

Figure 2.40: Sony BVW-35 Betacam SP® portable recorder/player. Courtesy Sony Communications Products Co.

Figure 2.41: Sony BVW-505 Betacam SP® Camcorder. Courtesy Sony Communications Products Co.

ing companies, local cable and low-power television systems, community colleges and wedding (or other special occasion) videography.

The VHS hi-fi format is the upgrade version of the familiar home consumer models, except the VTR circuitry is more advanced so the picture quality and hi-fi or stereo audio is reproduced with much greater precision. The rotary video head assembly (Figure 2.44) contains two video record/play heads, two audio record/play heads (right and left channel) and two dummy heads used for smooth tape handling during shuttle searches. Each complete pass of the video head assembly records a full field of video, so VHS is defined as a nonsegmented recording format.

The audio system consists of four separate high-quality channels with Dolby noise reduction (two hi-fi channels and two normal longitudinal channels). The hi-fi audio channels achieve their high quality by converting the amplitude modulated (AM) signals into frequency modulated (FM) signals and recording them onto the video track. This results in a relative tape speed of about 170 times the normal longitudinal audio recording speed. The VHS hi-fi format uses the ''depth multiplex'' recording system shown in Figure 2.45. With ''depth multiplex'' recording the hi-fi audio heads are displaced 120 degrees from, and mounted 50 micrometers higher than the video heads. In addition, the audio heads are angled at -30 degrees for the left channel and $+30$ degrees for the right channel, and the audio head gaps are about three times as wide as the video head gaps. The result of this displacement is the ability to record the audio deeper into the magnetic coating on the tape, followed by the video signal on the surface of the magnetic coating.

MII 1/2-inch VTRs

The MII 1/2-inch videotape format is the most recent entry into the video recording marketplace to take advantage of ultra-high-density recording methods. It is designed by Matsushita Electric Corp. of Japan and was premiered by Matsushita's U.S. distributor Panasonic Industrial Co. at the 1986 NAB convention. The actual EFP/ENG production VTRs began delivery in mid-1987, with NBC Broadcasting agreeing to convert essentially their entire television operation over to the MII 1/2-inch format.

MII VTRs use the component video recording and processing technology, along with chrominance time compression multiplexing (CTCM) and a new 1/2-inch metal-particle videotape in cassette form, to achieve a broadcast picture rivaling 1-inch quality video.

Basically, CTCM is the electronic process in which the luminance portion of the picture (Y) is frequency modulated (FM), while the chrominance portions (B-Y/R-Y) are time compressed, using delay circuitry, mixed together (multiplexed) and frequency modulated to form a separate chrominance channel. This process results in very little picture degradation throughout post-production and duping for broadcasting.

The use of metal-particle videotape allows up to 95 minutes of recording with the standard cassette or 23 minutes with the pocket-sized cassette.

Each VTR model features two longitudinal (AM) audio channels, two (FM) audio channels that are recorded in the video chrominance track and a dedicated longitudinal time code track.

CHOOSING THE BEST VIDEOTAPE FORMAT FOR A PROJECT

Although no one videotape format is right for every situation, the 1-inch helical formats have become extremely popular on many projects. This has been especially true since the three major broadcast networks switched to 1-inch type C—a move that prompted many major post-production facilities to follow suit. In 1986, NBC Television announced their plan for a complete changeover to Panasonic's new MII 1/2-inch format. It remains to be seen what type of impact this will have on the video industry.

Compared to the old 2-inch quad format, 1-inch resulted in lower equipment, tape stock and shipping costs. For example, 1-inch videotape stock was less than half the cost of comparable 2-inch videotape and required only half of the storage space. As other videotape format changes appear, the savings on stock costs and storage space for the new formats will probably provide even more advantages for the video producer's budget.

For the independent producer, there are several additional factors that must be taken into consideration. These factors include the location of the shoot, the equipment that is available for production, whether the offline workprint reels are going to be recorded simultaneously and whether the post-production facility to be used can handle the chosen format.

Electronic field production (EFP) shoots are usually recorded on 1/2-inch cassette, 3/4-inch cassette or 1-inch (B or C) format equipment. Relatively light, compact equipment is a necessity for any remote work that requires constant location changes. Hollywood director/cinematographer Richard Davies, for example, has used portable, smaller format equipment to

Figure 2.42: JVC BR-7700U VHS stereo hi-fi recorder/player. Courtesy JVC Co. of America.

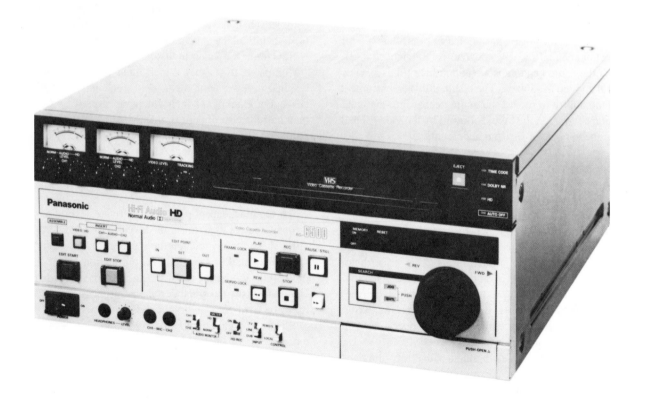

Figure 2.43: Panasonic AG-6500 VHS stereo hi-fi editing VCR. Courtesy Panasonic Industrial Co.

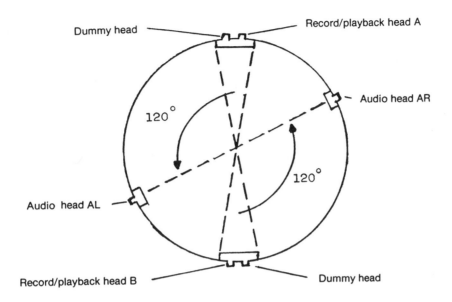

Figure 2.44: VHS hi-fi rotary video head assembly.

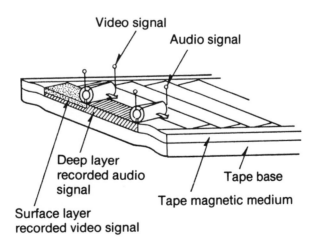

Figure 2.45: VHS hi-fi "depth multiplex" recording.

record video documentaries in many countries, including Red China and Uganda. He usually shoots on portable 1-inch VTRs or high-quality 3/4-inch or 1/2-inch Betacam® cassette recorders and then edits on 1-inch videotape. Davies reports significant savings on tape stock, equipment rental and shipping costs.

For in-studio production, the determining factor might be which equipment, if any, is installed at a particular studio. If equipment rental is necessary, cost and available editing formats could determine the choice. Be aware, however, that 1-inch type B tapes *will not* play on 1-inch type C videotape recorders, and vice versa.

The actual distribution format of a production need not be a deciding factor since copies from the edited master can be made onto the format chosen for distribution.

CONCLUSION

As I stated in the introduction to this chapter, anyone seriously considering a career as a video editor needs a working knowledge of the fundamental technical aspects of the industry. This chapter has presented the basic technical information needed to understand the video editing process. However, this has been only a primer and does not give detailed equipment specifications or detailed explanations of the many technical processes that are part of post-production. For readers who need or want to know more, the bibliography lists books that cover specific areas of concern in greater depth.

3 The Videotape Editing Bay

Editing bay configurations range from the simple and utilitarian to the posh and elegant, depending on the sophistication and the number of the system components. In fact, with so many different components available, choosing the right editing configuration may be one of the most important and difficult choices an editor makes. Regardless of whether he is buying or renting, an editor must select a system that is flexible and powerful enough to meet his current and projected post-production needs, while still being priced low enough to fit his budget.

Obviously, someone starting out on a shoestring should avoid investing in a full-blown computerized offline/online system—which could tie up more than $500,000 in equipment alone—unless there are solid project guarantees up front to support the system. In recent years, there have been several aspiring Hollywood editors who have made huge investments in state-of-the-art equipment without the necessary project guarantees, only to watch their facilities grow quickly outdated because they lacked the reserve capital needed to maintain and update the equipment. However, when shopping for an editing system, available capital should not be the sole factor in deciding on a particular configuration. After all, you're making an investment in business equipment, and you should expect that the investment will generate profits that can be put back into the system. If you lack this philosophy, you should probably forget about buying and simply rent someone else's equipment.

Editors on a shoestring budget might look into edit-ing configurations that grow as their needs grow. Several medium-priced editing controllers, such as the Videomedia, Inc., Z-6000 system and the Convergence Corp. ECS-195 unit, allow editors to upgrade the basic system as their needs change. Both systems can also offer either control-track editing or time-code editing capabilities.

The following sections describe these two types of editing and the advantages and disadvantages that each offers for the video editor.

CONTROL-TRACK EDITING SYSTEMS

Control-track editors (also called backspace editors and pulse count editors) perform edits by counting the frame pulses that are recorded on the control track signal (as discussed in Chapter 2). Like the editing controller shown in Figure 3.1, most control-track systems feature entry keys for entering edit-in and edit-out points, and many offer internal black video generators so editors can begin or end a program "in black." Of course, black video generators also permit preblacking master videotape stock.

Advantages of Control-Track Editing

The control-track system's biggest advantage is its relatively low rental or purchase price. For instance, the 1987 list price for a JVC Co. of America VET-3 control-track edit system equipped with two 3/4-inch VTRs—a system that permits only direct cuts between

Figure 3.1: JVC RM-86U control-track editing controller. Photo Courtesy of JVC Co. of America.

scenes—was around $7690, without video monitors. Control-track systems are also far less complicated to operate than time-code systems, since the editor merely aligns and enters the proper incoming and outgoing points on the playback and record VTRs and then pushes either the ''preview'' or ''record'' control. The editing system then backspaces both VTRs the proper number of control track pulses, rolls both VTRs so they are up to speed and performs the selected function.

Control-track systems also offer an inexpensive way of performing two-track audio editing, since the second audio track is not needed for recording time code. (More expensive videocassette recorders, such as Sony's BVU, feature a separate address track for recording time code, permitting both two-track audio editing and time-code editing on the same master tape.) With this in mind, an editing colleague of mine uses an inexpensive control-track edit system to piece together documentary footage, since it allows him to edit narration on audio track 1 while retaining all on-location sound effects on audio track 2.

One final advantage of using small-format, control-track editing systems is that you avoid the annoying

buzz, or cross talk, in the audio track that sometimes occurs during small-format time-code editing, particularly when the time code is recorded at too high a level during the production shoot.

Disadvantages of Control-Track Editing

The major disadvantages of control-track edit systems lie in frame accuracy, edit repeatability and, for those contemplating a control-track configuration as an offline system, the need for hand-logging the edit list. Unless the control-track system is kept in top shape, mechanical slippage can occur when the VTRs backspace and roll for an edit. This results in an error factor of plus or minus several frames, a factor that increases with each successive cueing. Once there is an inaccuracy, the edits are unrepeatable. An editor trying to repair a scene cut short by an early edit could wind up editing in reverse all the way to the beginning of the program!

The hand-logging of edit lists is covered in Chapter 5. Let me just mention, for now, that hand-logging edits requires up to 16 number entries per edit. Multiply that by the average number of edits in your proj-

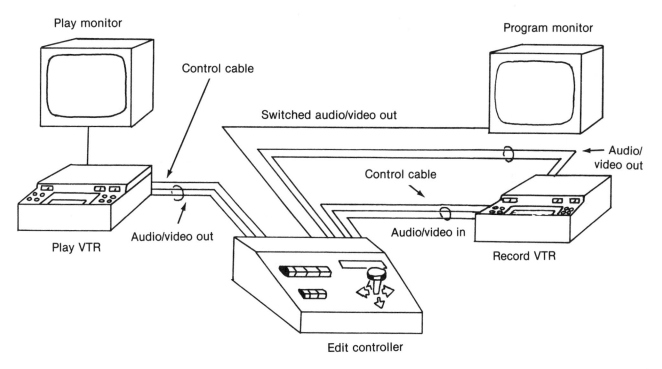

Play monitor

Control cable

Switched audio/video out

Program monitor

Audio/video out

Control cable

Audio/video out

Audio/video in

Play VTR

Record VTR

Edit controller

Figure 3.2: A two-VCR control-track editing system.

ects, and you get a pretty good idea of the human error factor involved—errors that don't show up until you enter the expensive online stage. In fact, the cost of trying to find several wrongly listed edits during online assembly can add up to the difference in the amount you would have paid for a time-code system.

As a rule, control-track configurations (as shown in Figure 3.2) rely on the automatic gain control (AGC) feature in the record VTR to control video levels during editing and on the VTR audio input controls to adjust audio levels. Consequently, there is virtually no control over the chroma saturation and phase characteristics of the video signal, nor is there any method of equalizing audio signals unless external equipment is added to the system. Adding a small audio mixer is relatively inexpensive, but the combined cost of adding a time base corrector, a waveform monitor and a vectorscope (items discussed later in this chapter) to adjust chroma and phase characteristics could be more than the cost of the entire control-track system. If you are using the control-track equipment as an offline system, the quality of the edit recording is a secondary consideration, and you probably won't need the extra components. However, if you plan to use control-track equipment in an online system intended to produce broadcast-quality edited masters, you had better count on investing in additional signal processing equipment.

AUTOMATIC TIME-CODE EDITING SYSTEMS

As an old mechanic once said to me in my hot rod days, "Speed costs money. How fast do you want to go?" The same applies to automatic time-code editing systems. Prices range from less than $10,000 to several hundred thousand dollars, depending on how fast and fancy your edits need to be. Anyone who has ever attended the National Association of Broadcasters annual convention can attest to the fact that there is an abundance of editing systems on the market, each with a variety of features and a myriad of optional add-on devices.

The basic versions allow cuts-only editing between two VTRs, with most including a line printer output port for printing the edit decision list. Add-on features include extra VTR playbacks, A/B roll editing with special effects generator (SEG) control for dissolves, edit list memory capacity, paper tape reader/punch or magnetic floppy disk drive units, software list management features, audio and video control interfaces, automatic assembly modes and so on.

Even the low-to-medium priced time-code edit controllers are quicker and more accurate than control-track editing systems, since they use the time code to search out scenes automatically. Some also permit internal wipes, dissolves and title keys, as well as accurate transitions with multiple-VTR systems and split edits in which the audio and video portions of the edit start at different times.

An important note for readers who are thinking of upgrading an existing system for use in offline editing: If you plan to improve your system so it is capable of generating edit decision lists, be sure that you first check with the post-production facilities where you are likely to perform your online assembly to make certain you are buying list-output equipment (such as reader/punches or floppy disk drives) and using software that are compatible with what is used at the online facilities.

Let's look at the EECO, Inc., IVES-II system as one example of a relatively low-cost time-code edit controller (see Figure 3.3). It is used primarily for electronic news gathering (ENG), cable TV operation, corporate video and educational and professional video. The EECO system can function as either a control-track editor or a time-code editor with two VTRs (one playback VTR and one record VTR). Time-code edits may be entered "on the fly" through the numerical keypad, or by frame-by-frame adjustment using the trim function.

With the IVES-II system, an internal color black generator allows variable rate fades from, or to, black. In addition, the editor can generate pre-blacked tapes recorded with time code by using the internal SMPTE time-code generator in conjunction with the color black generator. Existing source material videotapes can be coded (striped) for editing by using the STRIPE feature, while duplicate copies of tapes are made using the COPY feature. The system also features split edit capability and simple audio mixing from VTR or auxiliary inputs, plus a programmable general purpose interface (GPI) relay closure for triggering auxiliary equipment. EECO, Inc., also offers a list management interface to allow editors to output edit decision lists to a personal computer using the Edit Lister™ software program (see Chapter 5).

At a higher price level, there are the Videomedia Z6000 series systems and the Convergence ECS-200 series systems (see Figure 3.4), both of which are upgradable. In addition to the basic features offered by the EECO, Inc., IVES-II system, these two systems also include A/B roll editing from multiple source VTRs, internal edit list memory, auto assembly, complete internal list management capability (described in Chapter 5), video switcher control, and line printer and reader/punch or floppy disk outputs. Since they

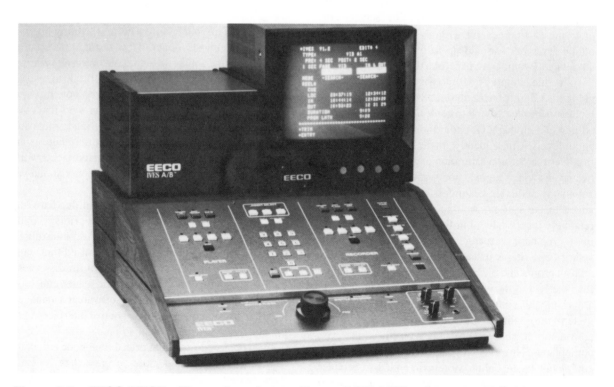

Figure 3.3: EECO IVES™ editing system shown with new IVES A/B™ enhancement allowing complete A/B control of three VCRs and special effects. Photo courtesy of EECO, Inc.

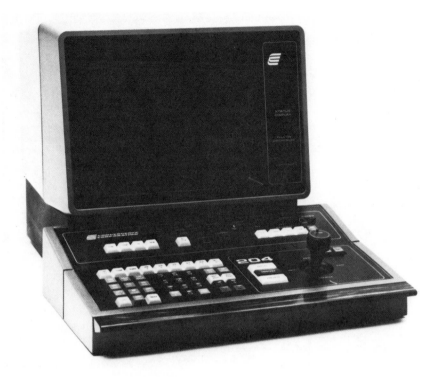

Figure 3.4: Convergence ECS-204 upgradable control-track and time-code editing system featuring auto assembly and edit list management. Photo courtesy of Convergence Corp.

can control multiple VTRs and feature auto assembly, the Videomedia and Convergence systems can be used for offline and online mastering. In fact, for budget-minded editors, these systems feature many of the same advantages found in the fully computerized edit systems, but at a significantly lower cost.

COMPUTERIZED TIME-CODE EDITING SYSTEMS

Computerized time-code editing systems are the most sophisticated of editing equipment systems. Figure 3.5 shows how the components of a computer edit system are interconnected and how signals travel to and from the individual pieces. On the pages that follow, you'll learn how the computer controls the system and how you, as a video editor, control the computer. Some of the other components are discussed in more detail in the section "Putting the Pieces Together," later in this chapter.

The Computer

The computer, the brains of the system, is actually a relatively small data processor comprised of three interrelated components: the central processing unit

(CPU), the internal real-time clock and the memory unit. The CPU is the central switching mechanism responsible for receiving and sending signals to and from the various pieces of equipment attached to the system. Synchronized to the house sync generator, the real-time clock performs the time computations that give the system its frame and time accuracy.

The third component, computer memory, stores the software to operate the system, as well as the edit decision list. The first computer editing systems had a memory capacity of 8K (about 8000 words), which seems small by today's standards. Current state-of-the-art edit system computers feature a memory capacity of 32K to 128K, with newer and larger systems appearing every year. In fact, the editing computers currently used at Hollywood's Unitel Video require 13K of memory just to load the software program that operates the system.

I should pause to point out that computer memories come in two types: volatile and nonvolatile. With volatile memory, the computer loses the data stored in memory whenever the system loses power. In contrast, nonvolatile memory systems store the data in magnetic cores that maintain their magnetism even when all power is lost to the system. Editors who are using a computer edit system with volatile memory should

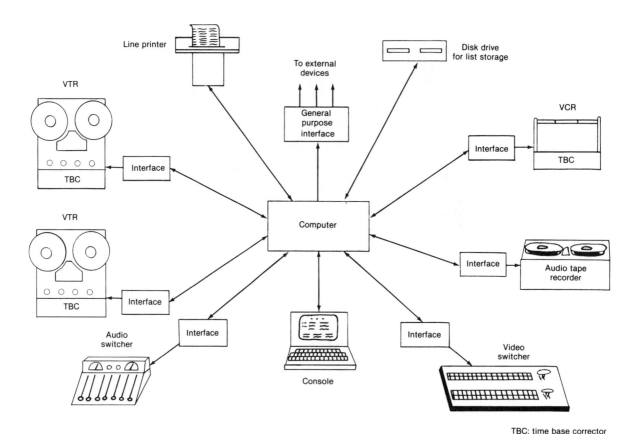

Figure 3.5: Input/output control signal path for a typical computer edit system.

think seriously about installing a backup battery so any stored data is saved if power to the system is accidentally cut off.

Some post-production facilities also install a microprocessor card, known as a memory map card, in their editing computers. Its purpose is to provide much faster input/output dialogue between the computer and the keyboard console and to soothe the nerves of editors who enter data through the keyboard faster than the computer could normally process it.

Interfaces

Interfaces are microprocessor devices that act as interpreters between the computer and the various types of equipment used in an editing system (VTRs, video switchers, digital effects generator, etc.). Since each type of equipment is designed differently, it stands to reason that a control signal that activates one piece of equipment will not necessarily activate another piece of equipment. Thus, in order to communicate, the computer first sends its own control signal to the interface board of the desired piece of

equipment. The interface board then changes the computer signal into the "right" signal for that particular component, allowing the computer to activate the component's function controls.

The Console

The console is the main terminal for controlling data input and output. It consists of a keyboard and a cathode ray tube (CRT) display unit. The keyboards of most major computer edit systems are based on the familiar ASCII (American Standard Code for Information Interchange) key format—the same format used on almost all small computer terminals. However, many edit system keyboards also include several "dedicated" keys for performing particular editing functions.

The CRT of most computer editing systems is a normal black-and-white video monitor on which is displayed the editing "menu," or list of available functions. (There are as many different menu displays as there are manufacturers of editing equipment.) Any data entered through the keyboard are also displayed on the CRT, as is the computer's return dialogue to the user.

BASIC EDITING TERMS

Before moving on to discuss a basic computer editing system, I should pause to define two terms that video editors use daily in their work: assemble edit and insert edit. These are the two fundamental types of edits used in all post-production sessions.

Assemble Edit

In assemble editing, new material completely replaces *all* information on the edit master tape from the point of edit. In other words, during an assemble edit, the record VTR's full-width erase and record heads are activated, and new video, audio and control track signals are recorded onto the master tape. This method of editing is very useful when an editor is piecing together long scenes such as talk show or game show segments, joining film-to-tape reel changes, or working with other projects that don't require editing into previously recorded material. Be aware, however, that if the new material needs to be edited into existing material (for example, inserting a cutaway shot into an existing dialogue scene), the record VTR must be in the insert edit mode. For reasons that are explained in the discussion of "insert edit" that follows, failure to heed this warning will result in severe video breakup on the out point of the edit.

Insert Edit

Insert editing was developed as a means of editing new material into previously recorded material without having video breakup at the edit-out points. For several reasons, it is the most common type of editing in use today. With insert edits, the VTR does not erase the existing control track. Instead, it locks into the control track recorded with the existing video material and permits editors to perform audio only, video only, or audio *and* video edits, depending on which erase and record heads the editor has activated. As a result, insert editing leaves the control track stable and continuous at all times.

Since small-format VTRs are especially vulnerable to unstable edit points, editors using such equipment often prefer using the insert edit mode. It is also advisable to use a time base corrector to feed the video signal to the record VTR. In this way, the video input

signal is extremely stable and, when used with insert editing, this technique greatly reduces problems of unstable edits.

Insert editing does have one minor drawback. Since the VTR must lock into an existing control track, the videotape stock being used for the edit master recording must be "preblacked." That is, before the edit session can begin, the videotape stock must be prerecorded, preferably with a continuous video signal containing a color reference burst (a black video signal, for instance), a continuous control track with frame pulses and continuous time code (if this is a time-code edit session) recorded at +2VU on the VTR audio meter.

Of course, most post-production facilities carry a reserve stock of preblacked edit master tape, or they can arrange to have it preblacked in advance of the scheduled edit session. As noted, some of the new small-format editing controllers have a preblacking feature built into the unit.

CONTROLLING THE BASIC COMPUTER EDITING SYSTEM

It is impossible to list detailed operating instructions for each of the top-of-the line computer edit systems, particularly since features and operating procedures differ significantly from one make and model to the next. There are, however, several basic features of computer edit systems that remain fairly constant from one manufacturer to the next and a number of standard functions that are found on most editing console keyboards. (A representative keyboard is shown in Figure 3.6.) Reviewing these basic functions and features should help aspiring editors gain a working knowledge of the computer-assisted videotape editing process.

VTR Selection

Each editing system must have a method for selecting and starting individual videotape recorders. In a basic cuts-only system, VTRs are usually selected by activating console switches labeled "record VTR" and "playback VTR." In more complex systems, as represented in Figure 3.6, the editor can choose from

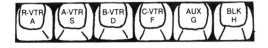

a row of keys labeled R-VTR (the record VTR) and A-VTR, B-VTR, etc., depending on how many

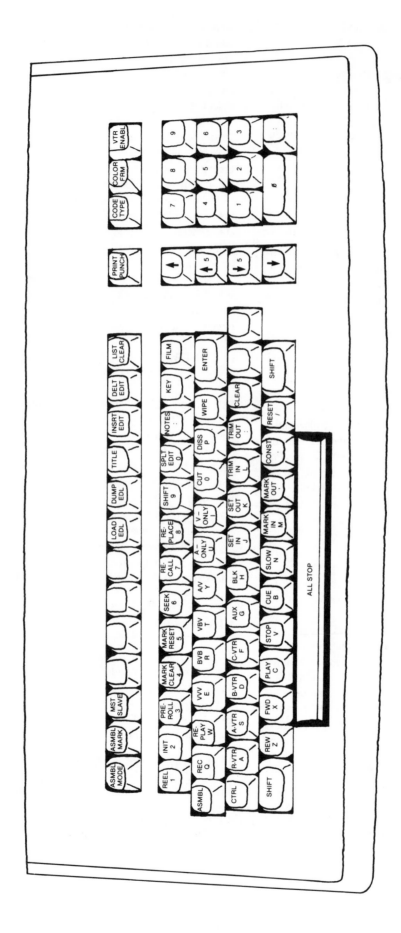

Figure 3.6: A representative keyboard layout for a computer edit system.

playback machines the system is capable of handling. All systems also feature auxiliary inputs for color bars, matte cameras, etc., and house reference black.

After an editor selects an individual VTR, he needs a way of controlling its functions. There may be remote controls for each machine near the edit console or a set of keys on the console that control the "rewind," "fast forward," "play" and "stop" functions. Other features include "all stop" (halts all VTRs in motion), "cue" (cues any VTR selected to a particular pre-entered time code location) and "slow" (puts the selected VTR in slow motion, usually to help locate an edit point). Some systems also feature a "jog" function (playback VTR frame advance or retard) for establishing video-to-audio track synchronization in the preview mode. Activating "jog" and pushing either the "advance" or "retard" control jogs the playback VTR one frame per button push, with the edit point entry number changed accordingly.

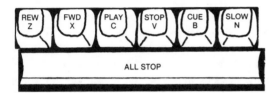

Edit Point Entry

Time-code edit systems, whether computer-assisted or not, employ three different methods of edit point entry. For the sake of simplicity, I will refer only to entering edit-in points. However, edit-out points are entered in precisely the same manner.

The first method enters an edit by pressing the "mark in" key while the selected VTR is either still-framed (for cassettes and helical VTRs) or in the playback mode (for all VTRs). This method is also called "on the fly" edit point entry.

In the second method, a predetermined edit point number is entered by pressing the "set in" key and then entering the eight-digit SMPTE time code corresponding to that edit point.

Once the edit point has been entered by either of these two methods and the edit has been previewed, an editor may find that it is necessary to adjust the edit points slightly. The third method, "trim in," allows the editor to do this by making plus or minus adjustments ("trims") to the edit point numbers. Trims may be entered in frame numbers or in hours-minutes-seconds-frames numbers, depending on the edit system.

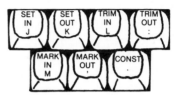

Multiple-Take Synchronizations Using the Trim Function

The trim function can be a very useful tool on complicated editing projects. For example, it often plays a key role in synchronizing multiple-take music tracks. Assume you are a director shooting a choreographed dance sequence on location. Working with a big budget, you have decided to shoot the sequence at 10 different locales. To make matters more complicated, you shoot three takes at each locale, giving the editor 30 different takes to synchronize with a master audio track. (This example assumes that the user bit time code, described in Chapter 2, is not being used.)

The editor first logs the sync start point for each selected take and then edits the master audio track onto the record master videotape. Once this is done, the editor has the sync start point for each take, plus the sync start point for the audio track on the edited master. With this information, the editor can determine the mathematical difference between the start point on the edit master reel and the start point on each take, simply by subtracting one code number from the other.

In the example used in Table 3.1, there is a 07:12:00:00 time code difference between take 1 and

Table 3.1: Time Code Numbers for a Sample Multiple-Take Synchronization Using the Trim Function

Take	Sync Point	Edited Master Sync	Difference (Offset)
1	08:32:10:00	01:20:10:00	07:12:00:00
2	08:45:15:10	01:20:10:00	07:25:05:10
3	08:50:25:20	01:20:10:00	07:30:15:20
4	09:02:31:00	01:20:10:00	07:42:21:00

the corresponding audio track point on the edit master reel, a 07:25:05:10 difference on take 2, and so on. Say that you, as the director, decide to cut to take 3 for four seconds at a point 30 seconds into the song (or 1:20:40:00). To do this, the editor would begin by entering the record start point (01204000) into both the record *and* the playback VTRs. Then, the editor would activate the ''trim in'' key for the playback VTR and enter the offset that has been calculated (07301520).

The resulting edit will occur from 01:20:40:00 to 01:20:44:00 (edit master time). The scene edited into the master will start at 08:50:55:20 and run to 08:50:59:20. Using this method, the editor will be able to maintain sync in a complicated situation with a minimum of effort. As a result, he will be able to spend his time creatively editing instead of performing endless computations.

Most sophisticated systems also employ one or more ''constant registers'' that are helpful in this sort of editing situation. Rather than repeatedly entering trim numbers in the eight-digit SMPTE code number, the calculated offsets may be stored in the constant register (one number per available register) and recalled by activating the ''trim in'' function and the constant key. This automatically recalls the number in the constant register, so the trim can be performed in one key stroke instead of eight.

Selecting Edit Modes

There are three basic edit modes: audio/video (A/V, also referred to as ''both''), audio-only and video-only. These are self-explanatory, and they are clearly indicated on any edit system keyboard. However, there are also several more sophisticated variations of these modes that require some explanation.

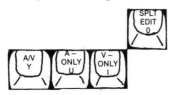

For example, many editing controllers permit split edits (provided that the record VTR is technically capable of performing this type of edit). In a split edit the audio and the video components of the edit begin or end at different times from each other, usually by a preselected time delay. For instance, an editor might use the split edit function to start the dialogue of one scene over the end of the preceding scene and then cut to the scene audio and video in sync *all in one*

edit function. I should emphasize that a split edit only *delays the start or stop point* of the audio or video portion of the edit. It does not *change* the audio and video sync relationship of the edit.

In another variation, editors using helical VTRs with dual audio track capabilities can perform A/V edits in which the audio edit is performed on track 1, track 2 or both, as well as audio-only edits on track 1, track 2 or both tracks. This is accomplished either by means of hardware (switches, patch cords, etc.) or by means of software (edit programs controlling electronic crosspoints), depending on the capabilities of the particular edit system in use.

Selecting Edit Transitions

There are four basic edit transition types: cuts, dissolves, wipes and keys. With computer editing systems, the editor enters instructions for performing transitions into the computer, and the computer does the rest. All edit systems are capable of performing cuts—transitions in which one shot or scene directly replaces its predecessor. This requires only the most basic of editing controllers to synchronize the two VTRs and to activate the record VTR. To perform dissolves, wipes and keys, you'll need an editing configuration that is more sophisticated. These systems include video and audio switchers as well as a time base corrector for each VTR.

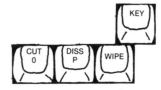

Dissolves

In a dissolve, the first scene fades out as the second fades in. In other words, a dissolve is a method of overlapping two shots for a transition effect. To perform a dissolve, the editing system needs to know which VTR contains the outgoing scene, which VTR contains the incoming scene, and how long (in frames) the dissolve will last. A typical set of computer instructions for performing a dissolve might read:

> From: A-VTR
> To: B-VTR
> Duration: 45

Wipes

Wipes are transitions in which a margin or border moves across the screen, gradually ''wiping'' out the first scene and revealing the second. A computerized editing system handles a wipe in much the same way that it handles a dissolve, with one exception. Since there are many types of wipes (horizontal, vertical, etc.), the editor must also tell the computer which wipe he wants. This is actually a fairly simple procedure, since every editing system comes with a chart illustrating the types of wipes it is capable of performing, along with a code number for each type. A typical set of instructions for a wipe sequence might read:

Wipe: #003
From: A-VTR
To: B-VTR
Duration: 45

Systems such as the CMX Corp. 340X also include delay dissolve and delay wipe features that allow the editor to cut in a scene and then dissolve or wipe after a predetermined time to another scene using only one edit entry. Other edit systems, including the Mach One manufactured by Fernseh, Inc., feature cluster edits that allow several edit functions (fade up, title key in/out, dissolve, wipe, fade out, etc.) to be performed in one edit entry.

Keys

Keys—special effects in which one shot is placed inside a portion of another shot—are the most exotic of the editing transitions. In professional video editing, keys are often used to insert titles into a scene or to perform sophisticated effects in which a character is placed against a background scene being generated by a playback VTR or some other video source. Of course, there are many variations from which the editor may choose, depending on the types of keys the system is capable of performing and the number of video sources available to the editor. A typical set of instructions given to the computer for performing a key might look like this:

Key in: Y
Background: A-VTR
Foreground: B-VTR
Delay: N
Fade: N
Duration: 45

For more information on performing keys, see the section on video switchers later in this chapter.

Preview Selection

All computerized editing systems allow editors to preview edits before actually performing them. In fact, most systems allow editors to choose from a variety of preview modes. Of course, the number and type of preview modes vary from system to system. I have described the most common types below.

Video-Video-Video (VVV)

The simplest and most popular type of preview is called video-video-video (VVV). In this type of preview, the system plays back the edited master reel to the entry point of the new edit, switches to the new scene that will be inserted (supplied, in most cases, from a reel on a playback VTR) and then switches back to the edited master playback at the edit-out point. In other words, by watching a monitor that is attached to the system, the editor can· see how the complete edit will look once it's performed.

Video-Black-Video (VBV)

Another preview mode available on most systems is video-black-video (VBV). In this mode, the video on the edited master cuts to black at the edit-in point, then back to video at the edit-out point. VBV previews are usually used to determine the proper pacing and length of an insert or to locate camera cut points when editing switchfed material.

Black-Video-Black (BVB)

A black-video-black (BVB) preview is the exact reverse of a VBV preview. In other words, the monitor screen will show black until the tape reaches the edit-in point, then video, then black again at the edit-out point. Usually, editors use the BVB mode to examine segments recorded on a source tape—segments that are being evaluated for possible inclusion on the edited master tape.

Engaging Manual Record

Once an edit has been previewed and any necessary edit-point adjustments have been made, it's time to perform the actual edit. In this section, I'll discuss the manual assembly mode; automatic assembly is described in the next section. Manual assembly is begun by activating the proper "record" key to trigger an operational sequence comprised of four steps: cue back, roll, synchronize and record.

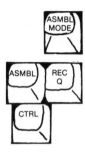

Once the "record" key is engaged, all VTRs involved with the particular edit in progress will cue their respective videotapes back a predetermined number of frames from the edit point. When all active videotapes are cued, or rewound to the proper start point, the computer issues a VTR "play" command. The VTRs then roll and "frame up," with the computer control system adjusting their speeds so they are synchronized and fully up-to-speed at the edit point. If the VTRs are not in exact frame-accurate synchronization at the edit point, the system will abort the edit and reset to attempt it all again.

Once the VTRs are correctly synchronized, the computer cues the record master VTR to go into record mode as soon as the record master tape reaches the time-code address selected by the editor. In addition, if the editor has entered an edit-out point, the computer system will trigger the record master VTR out of edit (out of the record mode) at the edit-out time-code address. If no edit duration (or a long edit duration) has been entered and the editor wishes to stop the edit prematurely, he merely activates the proper key and the record master VTR comes out of the edit cleanly.

Edit systems use different methods to activate the out-edit (or "stop edit") function. For example, the CMX 340X and the Interactive Systems Co. (ISC) Super Edit systems use the same key to start and stop the recording process, while the Fernseh Mach One system uses a "record" key for start and a separate "mark out" key for stop.

It is important to note that although the edit will begin on the first field of the frame marked in the "edit-in" time code entry, it will end at the last field *before* the frame marked in the "edit-out" time code. In other words, the edit will record up to, but will not include, the edit-out number. For example, let's assume that an edit has been performed from 01:01:00:00

to 01:01:15:00 on the edit master. The number for the next edit-in point should be 01:01:15:00 (not 01:01:15:01), since the previous edit actually ended on the field *preceding* the edit-out number.

Several editing systems also offer synchronized VTR roll and real-time edit functions. Of course, this has been possible for years by merely rolling all VTRs in sync and then camera cutting using a video switcher. However, the sync roll feature I'm describing here was originated by Fernseh, Inc. for the Mach One system. It allows editors to perform all switching between sources (cuts only) at the computer console in real time, while at the same time producing a frame-accurate edit decision list containing all camera cuts. This can be a powerful tool for an editor post-producing a project that was shot totally on isolated cameras. The system also offers a number of advantages for offline editing, including the ability to compile a quick camera-edited rough cut that can be further refined during successive fine cuts, while at the same time compiling a complete edit decision list for automatic online assembly.

Computer-Assisted Automatic Assembly

Automatic assembly requires a computerized editing system that is capable of storing numerous edit decisions in its memory and that can recall the decision data when needed. Of course, the number of edit decisions a system can store depends on the size of the computer's memory. The explanation of automatic assembly offered here is based on the operating procedures used with the CMX 340X computer editing system. Like many computerized editing systems, the 340X features two modes of operation: A mode (also called sequential or linear mode) and B mode (also called optimum or checkerboard mode).

A Mode

The A mode is a linear, sequentially programmed method of assembling edits. After the edit decision list is placed in the computer's memory, the editor enters the proper reel numbers for the tapes threaded on the playback VTRs and pushes the assembly key. When assembly is activated, unless otherwise instructed, the computer will start at the first edit and perform each edit in sequence until it arrives at an edit for which the proper reel is not currently assigned to a playback VTR. At that point the system will stop and indicate that another reel must be assigned. Once the correct reel is assigned, assembly can be reactivated. (The editor can also instruct the computer to skip that edit and proceed.)

This method of assembly works well for relatively straightforward projects that feature few production reel changes. It is also suited for projects that contain numerous effect transitions among a greater number of reels than can be assigned simultaneously on the editing system being used. (To make an effect transition, both source tapes must be available to the record VTR; see the section on match frame editing in Chapter 6.) Finally, editors who do not have access to a clean-up program for edit decision lists really don't have much of a choice. They *must* use A mode assembly, to avoid problems of edit overlaps.

B Mode

Of the two assembly modes, B mode assembly is the more efficient. In B mode, the computer system assembles an edited master by performing every edit on the edit decision list that can be completed using the reels currently assigned to playback VTRs. After those edits have been performed, the system halts and requests reel assignment changes. Consequently, edits are performed throughout the master reel, with spaces left to be filled by material selected from reels yet to be assigned.

Although it is efficient, B mode assembly does require that the following conditions be met before the session begins.

• *The edit decision list should be clean and accurate, with no edit overlaps.* When they must work with a decision list containing overlapping edits, editors should make sure that the new edits do not record over portions of previously performed edits. In addition, although relatively brief edit decision lists may be cleaned by the editor using the list management capabilities of the edit system, any complicated decision lists should be run through a computer with 409 clean-up capability. The 409 program (described in Chapter 5) automatically arranges the decision list for B mode assembly.

• *Decisions about the timing of edits should be made final before assembly begins.* If at all possible, editors should avoid making time changes in the decision list once assembly is in progress, particularly if those changes will affect previously performed edits. In most cases, the previously performed edits will have to be remade at the new edit times—changes that can result in a considerable waste of time and money.

• *If possible, the time code base placed on the edited master tape should be recorded on the edit VTR immediately before the assembly session.* In my experience, following this rule has resulted in far fewer editing problems related to bad tape stock and unstable edits. I usually inspect and evaluate the new tape stock prior to recording the video black and time code, right before each edit session.

• *The color setup of the production material should be properly adjusted before, and not readjusted during, assembly.* How completely you are able to adhere to this rule depends on how drastic the video and color imbalances become during the assembly session. Occasional, subtle changes in camera shots are best corrected after the assembly session when, upon viewing, they can be seen in the overall context. Trying to make minor color corrections during assembly almost always leads to overcorrection—a problem that causes much more trouble than the original, relatively minor imbalances.

Load and Dump Lists

All large computer editing systems feature memory devices for storing and loading edit decision lists, or EDLs. The method of loading and retrieving the EDL differs from system to system and depends, to a significant degree, on whether the system uses paper punch tape or magnetic floppy disk drives as its primary memory peripheral. Although many of the early computerized editing controllers used punch tape, floppy disks have become much more popular in recent years. Both are described in Chapter 5.

To transfer an EDL to and from memory, most editing computers employ some sort of "load list" or "dump list" key, as well as the always dangerous "list clear" key. For safety's sake, most computers also feature a fail-safe device that helps prevent editors from accidentally erasing an EDL they have spent hours or days compiling. For example, some editing computers are programmed to ask "Are you sure?" before they will dump, load or, in particular, clear a list. At the end of long, exhausting editing sessions, I strongly advise editors to stand up and stretch before they answer the computer's question. Giving the wrong answer has spelled disaster for a number of video editors—including the author.

PUTTING THE PIECES TOGETHER

For most people, the first visit to a computerized videotape editing bay can be both overwhelming and intimidating—something akin to being thrust into the cockpit of a space shuttle and launched. In my experience, the mistake most novices make is trying to absorb and comprehend the entire system at once.

Remember, first, that the whole is equal to the sum of its parts. It follows, then, that the total editing system is divided into several parts or components, each of which performs a different function. By separating the parts and examining their functions, aspiring editors can begin to get a handle on how the entire system works during the editing process.

We have already discussed the component at the heart of the editing bay: the computerized editing controller. In this section, we'll begin with an overview and then take a closer look at the other video and audio components that comprise a complete editing system.

Figure 3.7 shows how the video, audio and sync signals flow through the system. Notice that "house sync" (horizontal and vertical reference signals from the main sync generator) flows to every part of the editing bay concerned with generating or processing the video signal. This is necessary to maintain a common horizontal and vertical timing reference throughout the system.

Also, all video outputs from the playback VTRs (A, B, C), matte camera and digital special effects unit flow into separate input positions on the video switcher. As a result, each can be switched as a sepa-

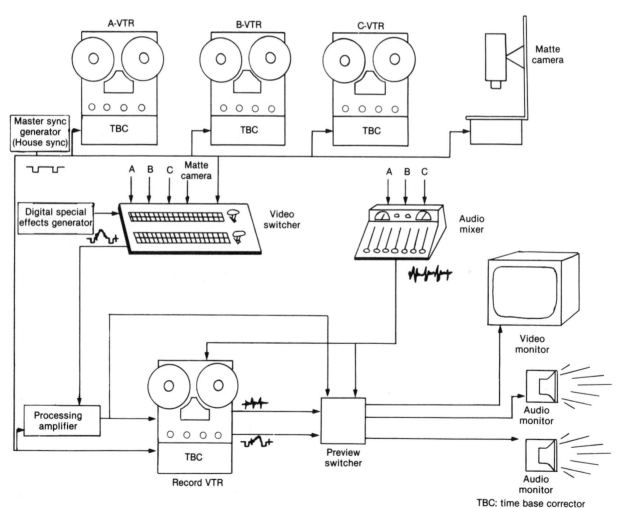

Figure 3.7: Signal flow through an editing system.

rate signal or can be combined with any of the other incoming signals to form switcher-generated special effects. Whichever combination of video signals the editor selects is then fed to a processing amplifier, where the sync pulse portions of the video signal are cleaned up to provide a very stable video signal. Finally, the stabilized video signal is sent to inputs on both the record VTR and the preview switcher.

The audio signals follow a similar path. After originating at playback VTRs or other audio sources, the sound signals are sent to the mixer board (the audio equivalent of the video switcher), where they are either switched individually or mixed with other input signals. The switched signal is then sent from the audio mixer to audio inputs on the record VTR and the preview switcher.

Notice that the video and audio outputs of the record VTR are also fed to the preview switcher inputs, instead of directly to preview monitors (as was the case with older videotape editing systems). The preview switcher is controlled by the computer, which tells it to switch from the record VTR audio and/or video playback (depending upon the mode of edit being performed) to the new audio and/or video signals at the desired edit point. The output of the preview switcher is then fed to the main audio and video monitors in the edit bay control room.

Videotape Recorders

The different VTR formats were described in Chapter 2. For now, I'll just mention again that all editing bays must be equipped with at least two VTRs: a playback or source VTR and a record or edit VTR. In actual practice, most computerized editing systems have multiple playback VTRs and at least one backup record VTR. In addition, many modern editing bays are equipped to record in multiple VTR formats (some combination of 2-inch, 1-inch, 3/4-inch and 1/2-inch recorders).

As Figure 3.7 shows, the playback VTRs send video source signals into the video switcher, which then sends the switched signal to the record VTR via the processing amplifier. Of course, like the other components in the system, the source and playback VTRs are all controlled by the editing computer.

Time Base Corrector

The time base corrector (TBC) is an electronic device that, when connected to the output of a VTR, corrects the stability and timing of the VTR playback video. It does this by stripping off the video signal's unstable horizontal and vertical sync pulses and replacing them with new, clean sync pulses. In addition, the TBC has controls for adjusting horizontal timing, pedestal (black), video level, chroma saturation and chroma hue.

Any multiple-VTR editing system that is capable of performing dissolves needs one TBC per playback VTR to ensure stable video transitions and edits. In addition, in small-format editing, TBCs provide a means of controlling video levels and ensuring a stable video signal.

The Video Switcher

Aside from the edit controller, the most elaborate component found in the editing control room is the video switcher. Actually, the switcher's primary function is quite simple: to feed the correct video signal into the record VTR. As Figure 3.7 shows, the video switcher comes between the source VTRs and the record VTR.

Figure 3.8 shows a fairly typical video switcher, the Grass Valley Group, Inc., model 1600-1A. Although the many buttons may seem confusing at first glance, they are actually arranged in four fairly straightforward rows: a program row, a preselect row and a pair of special effects rows (an A effects row and a B effects row).

The program row (PGM) is used to select the main output of the video switcher. By pushing the program row button for a particular video source, the editor would send the signal from that source to the record VTR. The preselect row (PST) is normally used to feed a selected signal to a preview monitor, so editors can preview a shot or set up an effect before sending it on to the record VTR as the program signal. The "take" bar allows editors to switch the signal from the preselect to the program row.

Effects Rows

Since special effects rows always come in matched pairs, the switcher shown in Figure 3.8 technically has only one full effects row. Using the effects buttons, editors can combine video signals from two or more sources to perform wipes, keys, mattes and other special effects.

For example, to perform a wipe manually, an editor would begin by selecting one video input on the A side of the effects row and another input for the B side. Then the editor would push the "wipe" button, select the desired wipe pattern from the patterns shown

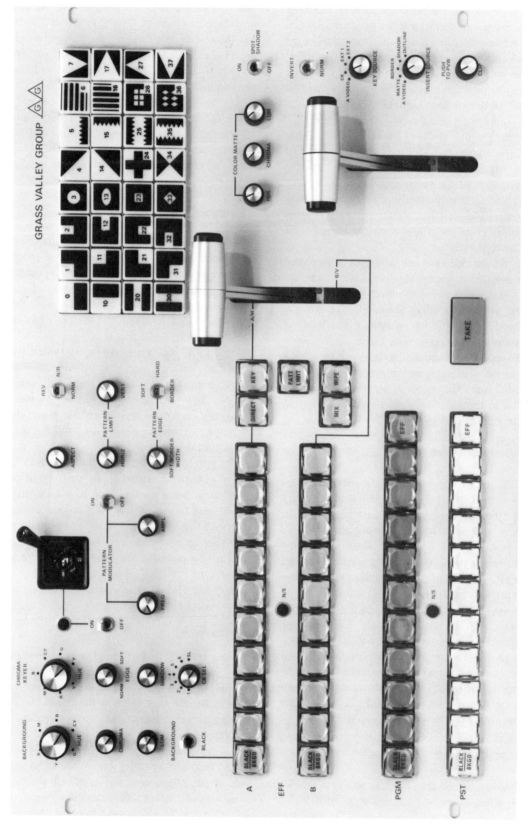

Figure 3.8: The Grass Valley Group model 1600-1A video switcher. Photo courtesy of the Grass Valley Group, Inc.

in the upper right corner of the switcher panel and move the fader handle from the A side to the B side. As the editor moved the handle, the system would perform the wipe. Of course, in fully computerized systems, the editing computer would perform the wipe functions automatically, following instructions stored on an edit decision list.

Key Effects

Keying is an almost equally straightforward procedure. For example, to key titles into a scene, the switcher operator selects the video background on the B side and the key (title) source on the A side and, after engaging the "key" button near the fader, moves the fader handle to the A side. This will key the title into the video background. To add some variety, the operator may want to put a border, shadow or outline around the title or to insert a color (matte) in the title.

The effect described in the previous paragraph is a relatively simple title key. Other, fancier key options require using other sorts of key and insert sources. Think of a key source as a hole cutter, and picture the insert source as the object filling, or inserted into, that hole. With that in mind, you should be able to see that using white objects of a selected shape as the key source has the effect of cutting a hole in the background video—a hole that is of the same shape as the white source. Then the editor or switcher operator can fill the hole with either a video picture (A video) or a solid color (matte) by setting the "insert source" switch. The resulting image contains the original background video, plus an inserted picture or color matte that is in the shape of the white objects. The switcher control marked "clip" allows the operator to adjust the video level of the key source so the keyed image is full and sharp.

The final type of key available on the GVG 1600-1A switcher is the chroma key. In this type of key, analogous to the blue screen technique used in film, a performer is set in front of a blue background that is then replaced by some other video source. To perform a chroma key on this switcher, the "key source" switch must be set in the CK position. The four chroma keyer controls ("hue," "edge," "shadow" and "CK sel") are in the upper left of the switcher panel. The "CK sel" knob selects the input RGB signal that feeds the chroma keyer (the keyer operates using only the red, green and blue primary colors of the key source video signal). As the swicher operator adjusts the "hue," "edge" and "shadow" controls, the background blue disappears, and another selected video signal takes its

place. The final effect is a single image in which a performer is "keyed over" a background picture supplied by the secondary video source.

I should mention one last feature on this GVG switcher—the background generator function. The background generator creates a full-color background signal that can be adjusted using the background "hue," "chroma" and "lum" controls in the upper left of the switcher panel. On this switcher, the background generator shares a panel position with the black signal generator. As a result, the operator must flip the switch to choose between "background" and 'black.''

Fancier Switchers

The GVG model 1600-1A we have discussed so far is a relatively simple switcher. Figure 3.9 shows the more sophisticated Grass Valley Group model 300-3A. Although it obviously offers many more features, this switcher operates on the same basic principles, with three effects rows (with additional video inputs), a "program" row and a "preselect" row on the lower panel.

In addition to these basic functions, the GVG 300 offers EMEM®, a computer memory device that allows complicated visual effects to be set up and stored for instant recall. Each effects row is also capable of dissolving or wiping background scenes while *simultaneously* performing a title key and a chroma key. The upper panel of the GVG 300-3A is actually a separate component that contains controls such as the Kaleidescope™ input selection and controls, and white pattern generator. (Digital video effects are discussed in Chapter 7.)

Waveform Monitor

Figure 3.10 shows a waveform monitor, a device used to display the output of the video switcher. The waveform monitor displays and permits measurement of the horizontal and vertical components of the composite video signal (described and discussed in Chapter 2). By expanding the sweep rate on the monitor, the editor can view either a "normal display" or individual lines of video. A normal horizontal display depicts a two-horizontal-line presentation of all lines of video information. The lines are superimposed on each other, with horizontal sync as the reference. This gives the editor a better overall view of video levels. If the editor prefers, he can also view the video signal devoid of all chroma, or he can view only the video chroma.

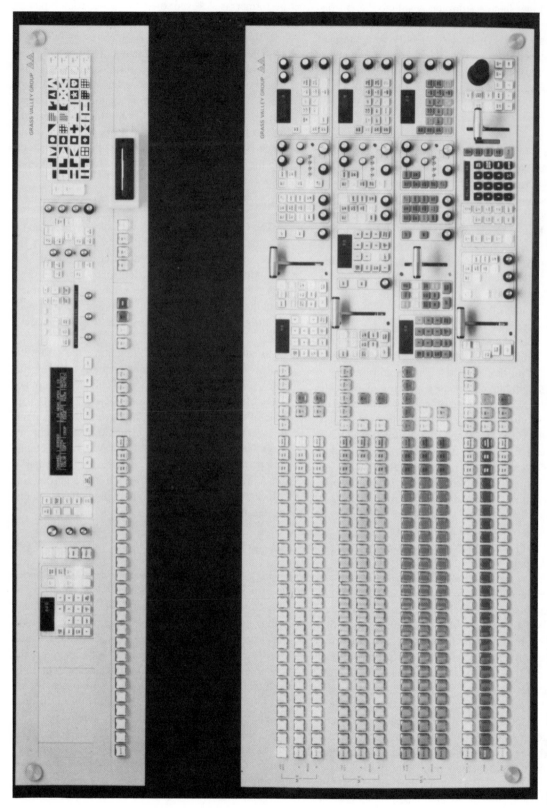

Figure 3.9:   The Grass Valley Group model 300-3A video switcher with Kaleidoscope™ digital effects system and effects memory system (EMEM®). Photo courtesy of the Grass Valley Group, Inc.

Figure 3.10: Tektronix 1480R waveform monitor for measuring the composite video waveform levels and timings. Photo courtesy of Tektronix, Inc.

Vectorscope

Another essential device associated with the video switcher is the vectorscope (see Figure 3.11). The vectorscope displays the color component of the video signal discussed in Chapter 2. During an editing session, editors must pay careful attention to the vectorscope display as they are setting up playback material.

The calibrated graticule on the vectorscope face indicates where the color bar test pattern colors should lie when properly adjusted. Adjusting the colors too low will cause the scene to have a pale, lifeless look (a desaturated picture). Adjusting the color too high will cause an oversaturated and noisy picture. (Figure 3.12 shows different vectorscope adjustments.) This is not to say that editors have no leeway in adjusting color levels. As a general rule, however, playback material should first be set up in the standard condition. Then, while viewing a properly adjusted color monitor, the editor can make whatever color corrections are necessary.

Video Monitors

Most modern editing bays feature a confusing variety of video monitors (see Figure 3.13). At most facilities, black-and-white monitors are used to display

Figure 3.11: Tektronix model 520 vectorscope for measuring the color components of the video signal. Photo courtesy of Tektronix, Inc.

Proper adjustment

Color adjusted too low (desaturated)

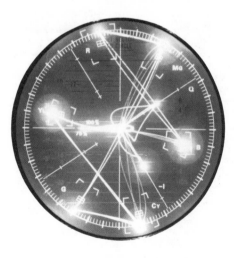

Color adjusted too high (oversaturated)

Figure 3.12: Vectorscope adjustments of color levels. Photos by Darrell R. Anderson.

the output of the various playback machines. Other monitors could show the output of each effects row of the video switcher, character-generated time code placed in the picture of the playback and record machines, or the output of any other video source being used in a particular editing situation.

There is also at least one large color monitor—the main program monitor—in each editing bay. All previews of edits and replays of completed edits are viewed on this monitor. In some editing bays, a second large color monitor is used to show the output of the preselect row of the video switcher.

Audio Mixing Board

Every editing control room contains at least one audio mixing board with equalization capabilities. Depending on the sophistication of the control room, the audio mixer may be single channel, dual channel or dual channel with stereo mixing capabilities (as is the mixing board shown in Figure 3.14).

Today, due to the increasing use of multiple audio track VTRs, dual-channel mixing boards are fast becoming a must for editing facilities. One point to remember is that while computer editing systems are said to control audio mixers as well as video switchers automatically, that control is limited to selection of the proper audio input crosspoints. Currently, the computer does *not* automatically control audio levels or equalization.

Volume Unit (VU) Meters

The sample editing bay we've been describing would have from one to three VU meters mounted next to the audio mixing board, depending on whether the board is single channel, dual channel or stereo. If the audio mixer is stereo, the meters will also indicate "left channel," "right channel" and "sum," as in Figure 3.14.

VU meters have a dynamic characteristic that approximates the response of the human ear. The VU meter, illustrated in Figure 3.15, measures the audio output of each mixer board channel in volume units, a unit of measurement designed to read between the average and peak levels of complex audio waveforms such as music or speech. The reference level established for VU measurement is referred to as zero level (or zero VU). This is equal to one milliwatt of power in the standard 600 ohm audio line.

Figure 3.13: Monitor display configuration in a computerized online editing bay. Photo courtesy of Vidtronics, Inc.

General Purpose Interface

Another component used in computer editing is the general purpose interface (GPI). The GPI is composed of several electronic switches that can be closed on commands from the edit control terminal. By telling the computer the proper switch and when to close it, the editor can start any type of electronic component (VTRs, audio mixers, etc.) that is capable of being remotely triggered. It is possible to close the switches before the edit point, at the edit point or after the edit point by entering the switch number and the desired SMPTE time code duration from the edit start point.

For example, a GPI could be used to start an audio tape playback two seconds before the edit point so it is up to speed when the edit is performed. To do this, the editor would tell the computer the proper switch position on the GPI, say position six in this case, and enter the duration "-2:00" in the position six register. On the other hand, to start the audio tape playback two seconds after the edit, the editor would enter the duration "2:00" in the proper register.

At most editing facilities, these small, sophisticated GPI units have replaced the large, clumsy panels of thumb wheel counters used in the early years of electronic editing. At Vidtronics, Inc., panels of thumb wheel counters (very similar to the EECO edit system counters described in Chapter 1) were connected to time code comparators that closed the switches when

the dialed-in time code number matched the time code signal fed to the counter panel. During the years before computer editing, we used these thumb wheel panels to control remote VTRs, video slow motion disk equipment and video switcher effects busses.

Title Cameras

Each online editing facility should have one or more black-and-white high-resolution title cameras. A post-production facility that intends to do television commercial work should have access to several. Ideally, the title camera should be mounted pointing straight down onto a copy stand that is equipped with adjustable lighting. Editors will find that the reverse polarity feature of high-resolution black-and-white cameras is invaluable for those clients who bring in black lettering on white cards. Offline edit systems can also benefit from having a title camera within easy reach for inserting titles in work-print copies that require client approval prior to on-line assembly. For a description of proper title card composition techniques, see Chapter 6.

Processing Amplifier

The processing amplifier, or proc amp, is used to generate and feed a stable composite video signal to the record VTR. It does this by taking the video output

Figure 3.14: Farrtronics, Ltd., stereo audio mixing board installed in a computerized online editing bay at Vidtronics, Inc. Photo by Darrell R. Anderson.

signal from the video switcher, separating the horizontal and vertical synchronization pulses from the signal, restabilizing them and then adding them back to the video signal.

The proc amp also features individual adjustments for sync level and timing, pedestal, overall video level, and chroma saturation and hue. In addition, it offers a clipper adjustment that editors can use to limit the video level to prevent overdriving the record circuitry of the record VTR. Proper clip level adjustment should be set at 100-104 IRE units. (Cable television services such as Home Box Office and Showtime require video levels that are no higher than 100 IRE units to prevent overloading when the program signals are sent through uplinks and downlinks during satellite transmission.)

Noise Reduction, Image Enhancement and Color Correction Equipment

Many editing facilities also feature a combination noise reduction, image enhancing and color correction system. By applying the various noise reduction options sparingly, bothersome ''noise'' in a video signal can be reduced appreciably. The reduction is espe-

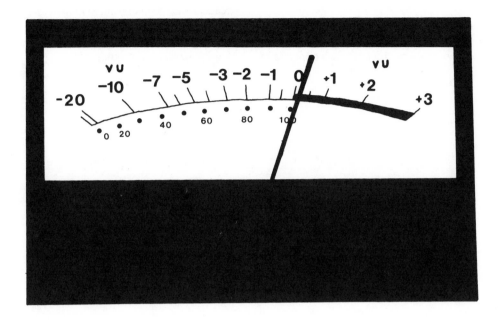

Figure 3.15: A VU meter.

cially noticeable in the black or dark areas of the picture, where noise shows up as gray, snowy streaks. However, using too heavy a hand on the noise reduction feature can degrade the picture by adding slight vertical impurities ("digitizing").

Image enhancement is the process of removing very fine detail lines from a picture while enhancing the more dominant detail lines, thus making the picture appear sharper and crisper. Enhancement can be used on the horizontal detail only, on the vertical detail only or on both. However, this is another case of "less is better." Overenhancement creates an artificial look in which large detail lines appear to have a "ringing" or halo effect.

The color correction feature allows editors to separate and make individual adjustments to the various primary and secondary color components of the video signal. In other words, unlike the hue controls on the time base corrector, VTR or proc amp, each of which changes the overall hue of the picture, the color corrector allows the editor to adjust individual components of the picture signal, much camera operator adjusts the color balance of a color TV camera.

The first step in color correction is to line the unit up at ground zero; that is, to adjust the VTR playback color bars for proper levels and phase with the color corrector in the "bypass" mode. Next, the editor ac-

tivates the color corrector and adjusts the various controls until the color bars appear normal in both the "correction" and "bypass" modes. The videotape is then shuttled to the area or scene needing correction, and the editor makes the necessary color adjustments. The problem may require only a small black balance adjustment in the red, green or blue areas, or it could require drastic alterations in gain and significant balance adjustments.

Digital Video Effects Controllers

Professional video post-production facilities of any size should have at least one digital effects system. There are several systems on the market, the four most popular of which are discussed in Chapter 7.

CONCLUSION

The review of the types of editing systems given at the opening of this chapter should help editors evaluate which of the types—control-track, automatic time-code or fully computerized time-code—is most appropriate for a particular project or for purchase. The right choice will depend not only on the available budget but also on the video format used and the complexity of the finished program as well as on whether the system is to perform online or offline editing.

This chapter also included a basic explanation of the operation of computerized editing systems and the functions they can perform. Finally, we've looked at the editing bay's components and their functions. For readers who need or want to know more specific information, Appendix B lists several available video editing systems; manufacturers' addresses and phone numbers are given in Appendix C.

4 Preparing for Post-Production

Consider the following scene from an actual editing project. The producer and director arrive at the post-production facility after their day of shooting, displaying great impatience to begin. After greeting them, the video editor notices that they do not appear to have brought their original production tapes. When he asks where the original material is, the producer replies in all seriousness, "I don't know. Do we need it?"

As incredible as this true account sounds, it points out one of the biggest obstacles to successful video post-production: the lack of proper preparation. More than anything else, achieving professional results in video editing is a matter of tireless attention to detail. Video post-production is, in effect, a manufacturing process and thus involves a logical sequence of events, as shown in Figure 4.1. These steps are discussed in Chapters 5 and 6, but before the process can begin, there is a great deal of organizational work to be done.

Because videotape projects generally have short delivery deadlines and narrow budgets, inadequate preparation inevitably leads to one or both of the following: cost overruns due to overtime expenses for facilities and personnel or an inferior product due to insufficient time to edit the project correctly.

Rates for final assembly editing on computerized systems have reached several hundred dollars per hour. To make the most of their editing time, producers or editors in charge of a post-production session should check the editing facility beforehand. Here are some things to look for.

• After determining the special effects that need to be added during post-production, find out if the editing facility is equipped to perform them. With new digital effects devices coming onto the market in ever increasing numbers, no single post-production facility could possibly have all the equipment needed for every editing situation.

• Find out which videotape formats the editing bay is equipped to handle and whether you will need to make format conversion duplicates (also called bump-ups).

• If a project that is to be edited offline contains several complex special effects, find out if the post-production facility will schedule a "pre-ops" online session in which the effects are performed and edited into the workprint reels. This is particularly important if producers or network executives must approve workprints before final online editing.

In addition to these basic determinations, preparation for post-production includes: developing budgets, reserving post-production facilities, selecting videotape stock, completing film-to-tape transfers and organizing an editing team. These activities are described in this chapter.

PREPARING A BUDGET

Developing an accurate, detailed budget is the single most important task in preparing for post-production.

To prepare an accurate budget, editors and producers must begin by determining the project's exact post-production requirements. Then they can go on to specify what editing and effects equipment they will need and the length of time for which the equipment and editing facilities will need to be scheduled.

Table 4.1 is a sample budget form for videotape post-production, and in this section we will look at each item included in the table. Since each editing project is different and since many projects don't require the "full treatment," I have broken the budget form into five sections: pre-duping costs, offline costs, online costs, audio sweetening costs and special costs. Once you determine which costs and categories apply to a particular project, you can fill in the appropriate blanks.

Pre-Duping Costs

Pre-duping includes those items needed in advance of the editing session. Window dupes, for instance, will be required for offline editing of any material shot on location or of any other material for which window dupe copies were not recorded during production. Format conversion will be necessary if the editing facility cannot handle the format of the original production tapes. As described in Chapter 2, time coding is the process of recording SMPTE time code on the appropriate audio track of the VTR.

Submaster duping may be required if two copies of the original material are needed or if a protection copy is desired. Audio-to-VTR transfer is self-explanatory and includes any audio tapes or records that must be transferred to videotape prior to the edit session. Film-to-tape transfer refers to any straight across, pan-scanned or color-corrected transfers needed for the actual editing session. (A more detailed discussion of film-to-tape transfer appears later in this chapter.) The last item in this category covers any additional videotape stock costs not included as part of previous items.

Offline Editing Costs

The second budget category includes those costs that pertain to the offline editing process. The cost for the offline editing system itself is self-explanatory,

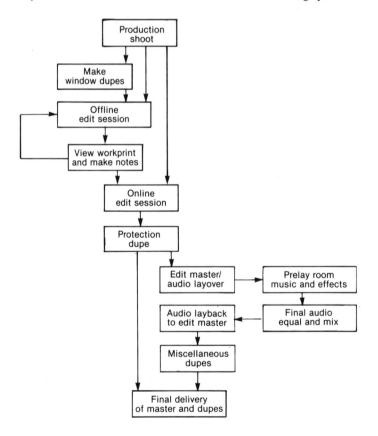

Figure 4.1: Post-production flowchart

Table 4.1: Sample Post-Production Budget

Proposed Budget for _____

Pre-Duping Costs

Type of Services and Equipment Used	Rate Per Hour	X	Hours of Use	=	Total Cost
Window dupes	_____		_____		_____
Format conversion	_____		_____		_____
Time coding	_____		_____		_____
Submaster duping	_____		_____		_____
Audio-to-VTR transfer	_____		_____		_____
Film-to-tape transfer	_____		_____		_____
Additional tape stock	_____		_____		_____
				Subtotal	_____

Offline Editing Costs

Type of Services and Equipment Used	Rate per Hour	X	Hours of Use	=	Total
Offline editing system	_____		_____		_____
Preblacked master	_____		_____		_____
Videotape stock	_____		_____		_____
Title camera	_____		_____		_____
Computerized edit decision list cleanup	_____		_____		_____
Overtime labor	_____		_____		_____
Dupes of workprint	_____		_____		_____
				Subtotal	_____

Online Editing Costs

Type of Service and Facilities Used	Rate per Hour	X	Hours of Use	=	Total
Online editing system	_____		_____		_____
Title camera(s)	_____		_____		_____
Color camera(s)	_____		_____		_____
Electronic graphics	_____		_____		_____
Digital effects	_____		_____		_____
Portable audio equipment	_____		_____		_____
Noise reducer/color correction	_____		_____		_____
Preblacked master videotape stock	_____		_____		_____
Extra personnel	_____		_____		_____
Overtime labor	_____		_____		_____
Duping	_____		_____		_____
Extra videotape stock	_____		_____		_____
Cassette or VHS dupe stock	_____		_____		_____
				Subtotal	_____

Audio Post-Production Costs

Type of Equipment and Facilities Used	Rate per Hour	X	Hours of Use	=	Total
Layover	_____		_____		_____
Sound effects editing	_____		_____		_____
Audience reaction system	_____		_____		_____
ADR, foley and announce	_____		_____		_____
Final mixing room	_____		_____		_____
Layback	_____		_____		_____
Extra personnel	_____		_____		_____
Overtime labor	_____		_____		_____
Multitrack audio stock	_____		_____		_____
Miscellaneous audio stock	_____		_____		_____
				Subtotal	_____

Special Costs

Type of Equipment and Facilities Used	Rate per Hour	X	Hours of Use	=	Total
_____	_____		_____		_____
_____	_____		_____		_____
_____	_____				_____
			Subtotal		_____
			Total of subtotals		_____
			Sales tax (if applicable)		_____
			Total budget		_____

as is the charge for preblacked master videotape stock. (There is usually an extra charge for preblacked stock, as opposed to blank videotape stock; hence the separate listing.) Title cameras may be included with the editing system or cost extra, as may the cost of cleaning up the computerized edit decision list (EDL). This last item includes the cost of cleaning and tracing a computer EDL (discussed in Chapter 5) or of generating a new edit list from hand-logged EDLs compiled on a non-time-code offline system. Overtime labor costs are also self-explanatory but can fluctuate greatly depending on the efficiency of the post-production schedule. The last item in this section covers workprint duping costs that are incurred if copies need to be sent to producers or network executives for approval.

Online Editing Costs

The third section, and the most complex, covers online editing costs. There is, of course, the cost of the online editing system itself, which increases with the number of playback VTRs the project uses. I budget and schedule the playback VTRs by determining how many reels of original material there are, plus how frequently the reels will have to be changed during the session. For example, if a project has 20 hours of original material but the tapes run in sequence, I wouldn't schedule four playback VTRs, since two or three would sit idle for most of the session. On the other hand, if the project has only six to eight reels of material but the session calls for special effect transitions and frequent reel changes, I would schedule at least three playback VTRs to avoid wasting time waiting for reel changes.

Although many post-production facilities include one title camera in the base price of the edit system, most charge extra for additional cameras. The extra charge for adding a color camera is self-explanatory.

I have included the cost of electronic graphics in the online category. However, most electronic graphics work should actually be completed before the online session, to prevent expensive delays. Some facilities charge by the hour, while others charge a flat rate for the job. Find out which is the case for the editing facility you've chosen.

Digital effects are generally billed as a separate cost. They can be voracious budget eaters, ranging from a base cost of $150 per hour to $1200 per hour for multichannel systems.

The portable audio equipment category includes the cost of audio transfers during online editing (although many facilities include this charge in their basic edit system price). Noise reduction, image enhancement and color correction (described in Chapter 3) complete the list of extra equipment charges for online editing.

As was the case in the offline category, preblacked master costs are listed separately from blank tape stock, since there is usually an extra charge to preblack the tape. Depending on the complexity of the project, there may also be charges for additional personnel and overtime. Duping charges cover the cost of making a protection copy of the final edit master and of making any additional small-format copies. (By the way, it's usually more economical to make all the dupes simultaneously.) The final two items in the online category cover any extra charges for videotape stock incurred during the online session.

Audio Post-Production Costs

The audio post-production category begins with the layover charge—the fee that includes the cost of transferring the audio tracks and time code from the edited master videotape onto a multitrack audio recorder. A workprint copy of the video is made at the same time and is used as a visual reference during the post-production process.

The sound effects editing charge covers the cost of renting the small audio room away from the main mixing room where sound effects, dialogue, music or audience reaction tracks are added to the multitrack recording. Many stage production projects will rent a special system to add the audience reactions to the multitrack recording.

The ADR, foley and announce section refers to the fee for the recording stage used for recording Automatic Dialogue Replacement (ADR), foley sound effects and any additional narration tracks needed during sound effects editing. However, the cost of the talent would be considered a production cost. The final mixing room is used after all music, dialogue, narration and effects tracks have been edited into the multitrack recording, and the recording is ready for mixing and equalization.

The layback charge is the fee for recording the final mixed audio tracks back onto the edit master videotape (synchronized, of course, with the video track). Once again, we conclude with slots for extra personnel, overtime and tape stock costs.

Special Costs

This final section is a miscellaneous category for adding in those last-minute items not accounted for in the main budget. With experience, the producer and editor will learn how much money to set aside for the inevitable extras.

RESERVING POST-PRODUCTION FACILITIES

Editors and producers must be careful to plan for and reserve sufficient editing time at the post-production facility. Editing equipment is generally reserved by the hour, and it is normally in high demand. Reserving insufficient time results either in not completing the project or in making the facility's next client wait—a decision that is made by the operators of the facility, not by the editor or producer.

Generally, it is a good idea to reserve 30%-50% more time than the project should reasonably take to finish. Most reputable post-production facilities

charge only for the time used, not the time initially reserved. Naturally, the facility's policy should be confirmed beforehand.

When reserving equipment time, be sure to give all the information needed to schedule your project properly. It's not at all uncommon for clients to show up for a scheduled edit session with film or "mag" audio tracks that need to be transferred but that were never mentioned when the project was originally scheduled. This "oh, by the way" syndrome creates real problems if the facility's equipment is tightly booked. Of course, conditions and needs change throughout the course of a project, and last minute additions are a way of life in the post-production business. However, whenever possible, try to alert the facility about any change in equipment requirements before arriving for the session.

Table 4.2: Post-Production Checklist

Title of project _____

Approximate length of project _____

Projected completion date _____

Client's name _____   Client's phone number _____

Video equipment specifications

Format of original material
 (2-inch, 1-inch, etc.) _____
Video equipment required:
 VTRs (playback and record) _____
 Digital effects equipment _____
 Title camera(s) _____
 Color camera(s) _____
 Crawl machine _____
 Electronic graphics equipment _____
Time code type _____

Telecine specifications

Film type (neg, IP, etc.) _____
8mm, 16mm or 35mm? _____
MOS, optical or mag sound? _____
Color correction required? _____
Pan scan? _____
Slide conversions _____
Videotape conversion format _____
Time code type _____

Audio specifications

Videotape format _____
1/4-inch, 2-track or multitrack
 audio required? _____
Sound effects? _____
Music? _____
Turntable required? _____
Announce booth required? _____

VTR duplication specifications

Number of dupes _____
Videotape format of master _____
Videotape format of dupes _____
Master will arrive/already at facility _____
New tape stock required/client
 stock provided _____
Color bars at head of dupe/
 between spots _____
Dupes recorded on separate reels? _____
Other special requirements _____

Offline workprint duplication

Videotape format _____
Time code on audio/address track? _____
Time code characters in picture? _____

Miscellaneous

Hours reserved at editing facility _____
Name and phone number to
 contact _____
Client contacted? _____
Original material will be shipped in
 advance? _____
Special shipping information _____
Billing instructions _____

To help editors and producers reserve the right equipment for the right amount of time, I have drawn up a post-production checklist (see Table 4.2). Naturally, every item on the list does not apply to every editing project.

USING YOUR OWN VIDEOTAPE STOCK

Some production companies prefer to purchase their own videotape stock for recording the edited master tape. Although there is nothing wrong with this, the executives in charge of post-production should remember that, since virtually all master editing is done on preblacked, time-coded "base" tape, the production company must deliver the stock to the post-production facility prior to the editing session. Keep in mind that evaluating, preblacking and time-coding the tape is a real-time process. That is, it takes one hour to record "base" on a one-hour reel (actually, more like one and one-half hours for a one-hour tape, when you include evaluation and setup time).

I should also point out that production companies supplying their own videotape stock for the edited master bear full responsibility for the quality of that stock. If, after several hours of editing, it is discovered that the edit master stock is defective, the production company may have to pay for several extra hours of online editing time.

Choosing the right videotape stock is not an easy task, since all manufacturers naturally claim that their product is superior. For production companies that lack the equipment needed to evaluate tape stock, I would recommend checking several post-production facilities. Find out what stock they prefer and why. Ask about the frequency of drop outs (indicating impurities in the oxide layer), video noise content, audio frequency response and the way the tape winds around the reel hub (improper winding can cause edge damage and tape cinching).

For those production companies that own VTR equipment with test monitors, I recommend getting tape samples from each manufacturer and performing the following test. Start by recording color bars and the normal 0 VU 400 Hz audio tone on the tape for two minutes. Then, record a gray scale (no chroma) video signal and a 1 kHz audio signal at 0 VU for another two minutes. Repeat this process in at least five locations on each reel.

Now set up the VTR for proper playback levels and replay each test recording. While watching the color monitor and audio meter, count the number of drop outs (random lines or streaks in the picture) and the amount of audio deviation about the 0 VU mark on the VU meter. (See the VU meter illustrated in Figure 3.15, Chapter 3.) Then test for video noise by playing back the gray scale video recording as you watch a vertically expanded horizontal waveform presentation. Note the amount of "fuzziness" (video noise) on the lines of the gray scale signal. Finally, run one more audio frequency response check by measuring the amount of deviation around the 0 VU mark of the VU meter with the 1 kHz tone playback. By comparing the test results from different types of videotape stock, you will end up with a pretty fair idea of which videotape brand you should buy.

PREPARING FILM-TO-TAPE TRANSFERS

Post-production often involves using material that was originally shot on film and that must be transferred to videotape. This film-to-tape transfer raises a number of complex questions:

• How is a film, shot at 24 frames per second, transferred to a video recording at 30 frames per second?

• How are film footage counts converted to SMPTE time code?

• How do editors set up proper film "A/B roll transfers" to videotape?

This section addresses these questions and offers some practical suggestions for setting up and performing film-to-tape transfers.

Transferring 24-Frame Film to 30-Frame Television

As you know, the standard U.S. television signal operates at 30 frames per second. Each television frame is, in turn, composed of two television fields, making for a TV signal scan rate of 60 fields per second. (As mentioned in Chapter 2, the actual scan rate is slightly lower: 59.94 Hz to be precise.) Since film operates at a rate of 24 frames per second, we must find six extra frames per second somewhere to make up the difference during a film-to-tape transfer.

This problem was initially solved several years ago by installing a mechanical device known as an intermittent pull-down assembly in the film projector. The pull-down assembly works on the principle of periodically holding a film frame in the projector gate for three television fields duration (one and one-half TV frames) instead of the normal two fields duration (one

TV frame). Keep in mind that 6 television frames equal 12 television fields. It follows, then, that by holding every other film frame in the gate for an extra field duration, the pull-down device produces the 12 extra fields, or the 6 extra frames, needed to make up the difference between film and TV frame rates.

This is how it works. The pull-down device holds the first film frame in the projector gate for two TV fields (one TV frame). Then, the second film frame is held in the projection gate for three television fields. As a result, the second television frame consists of two fields of the second film frame, and television frame three consists of one field of film frame two and one field of film frame three. Table 4.3 illustrates the pull-down ratio of film frames to TV frames during the film-to-tape conversion process—and how it all adds up to 24 film frames filling 30 TV frames each second.

The pull-down process can present problems during editing. For example, if a pull-down occurs in conjunction with a camera cut, the particular television frame involved would contain one field of the outgoing scene and one field of the incoming scene. If this film-to-tape transfer is then edited on videotape, an edit performed on that particular frame will appear as an annoying "flash frame." Fortunately, this can be corrected by adjusting the edit point one frame. Also, in a "pan scan" operation, the pan will appear as a double image or miscut if the pan starts on a pull-down frame. This can be remedied by advancing or retarding the pan start by one frame.

Using Flying Spot Scanner Projection Systems

Today, many post-production facilities use flying spot scanner systems to transfer film to videotape (see Figure 4.2 and Figure 4.3). Unlike the mechanical pull-down systems, flying spot scanners use a continuously moving electron beam to scan the film as it rolls. As a result, there is much less stress on the film during the transfer process.

With the increasing use of flying spot scanner systems, it's becoming much more common to transfer internegatives, rather than lab prints, for subsequent editing on videotape. The internegatives are color corrected during the transfer process, using electronic colorimetry correction systems. This saves both time and film processing fees and results in a high-quality color-corrected videotape for use as either an editing workprint or a completed master.

Two frequently used options available for flying spot scanner systems offer particularly useful

Table 4.3: Pull-Down Rates in Film-to-Tape Transfer

Television Frame	Film Frames In Field 1	Film Frames In Field 2
1	1	1
2	2	2
3	2	3
4	3	4
5	4	4
6	5	5
7	6	6
8	6	7
9	7	8
10	8	8
11	9	9
12	10	10
13	10	11
14	11	12
15	12	12
16	13	13
17	14	14
18	14	15
19	15	16
20	16	16
21	17	17
22	18	18
23	18	19
24	19	20
25	20	20
26	21	21
27	22	22
28	22	23
29	23	24
30	24	24

capabilities. The variable speed option makes it possible to vary the film transfer rate between 16 and 30 frames per second. Theatrical features sold for TV broadcasting are sometimes transferred at a faster than normal rate so the films will fit into TV time slots, without cutting scenes or segments. For example, when transferred at a rate of 28 frames per second, a feature film that normally runs for 115 minutes would end up taking only 98.5 minutes—a difference of 16.5 minutes, without eliminating one frame of film. The zoom option available on some scanning systems makes it possible to isolate specific areas within the film frame during the transfer process.

Film Transfers Using a Telecine Synchronization System

Today's filmmakers not only transfer original negatives or prints to videotape for post-production, but also synchronize the original 1/4-inch production audiotape with the film during the transfer. Telecine synchronization systems such as the Time Logic Controller (TLC), plus the new time code Nagra audio

Figure 4.2: The MKIIIC Rank Cintel Flying Spot Scanner Telecine. Photo courtesy of Rank Cintel Ltd.

Figure 4.3: Rank Cintel remote control panel for film-to-tape transfer. Photo courtesy of Vidtronics, Inc.

recorders used during the production, have allowed filmmakers to eliminate the intermediate step of creating magnetic film sound transfers and synchronizing them with the negative.

The TLC operates much like a video editing system, except that it controls one or more flying spot scanner telecines, a video production switcher, and any combination of up to four video or audio recorder/players. TLC works with either NTSC or PAL and has the ability to perform, assemble or insert edits; control the video production switcher to cut, dissolve, wipe or split A/V edits with field accuracy; and automatically calculates and performs variable speed transfers. Telecine synchronization systems have now opened up the possibility of creating an edited, color corrected master videotape

directly from the original film negative.

Other important features include the ability to simultaneously create up to four different delivery formats of a transferred film program. Its one-field accuracy enables the user to replace damaged or problematic shots in the previously transferred footage. For example, wide screen features that are pan-scanned for television must have the title sequences redone and inserted to fit the television aspect ratio. Systems like the Time Logic Controller make this an easy process.

Converting Film Footage to Television Frames

Video editors often need to know how long a given length of film will run once it is transferred to vid-

Table 4.4: Conversion of Film Footage to TV Time for 16mm Film

| 16mm Film | | Television | | | | 16mm Film | | Television | | | |
Feet	Frames	Hours	Minutes	Seconds	Frames	Feet	Frames	Hours	Minutes	Seconds	Frames
0	0	0	0	0	0	0	38	0	0	1	17
0	1	0	0	0	1	0	39	0	0	1	19
0	2	0	0	0	2	1	0	0	0	1	20
0	3	0	0	0	4	2	0	0	0	3	10
0	4	0	0	0	5	3	0	0	0	4	30
0	5	0	0	0	6	4	0	0	0	6	20
0	6	0	0	0	7	5	0	0	0	8	10
0	7	0	0	0	9	6	0	0	0	9	30
0	8	0	0	0	10	7	0	0	0	11	20
0	9	0	0	0	11	8	0	0	0	13	10
0	10	0	0	0	12	9	0	0	0	14	30
0	11	0	0	0	14	10	0	0	0	16	20
0	12	0	0	0	15	11	0	0	0	18	9
0	13	0	0	0	16	12	0	0	0	19	29
0	14	0	0	0	17	13	0	0	0	21	19
0	15	0	0	0	19	14	0	0	0	23	9
0	16	0	0	0	20	15	0	0	0	24	29
0	17	0	0	0	21	16	0	0	0	26	19
0	18	0	0	0	22	17	0	0	0	28	9
0	19	0	0	0	24	18	0	0	0	29	29
0	20	0	0	0	25	19	0	0	0	31	19
0	21	0	0	0	26	20	0	0	0	33	9
0	22	0	0	0	27	21	0	0	0	34	29
0	23	0	0	0	29	22	0	0	0	36	19
0	24	0	0	0	30	23	0	0	0	38	9
0	25	0	0	1	1	24	0	0	0	39	29
0	26	0	0	1	2	25	0	0	0	41	19
0	27	0	0	1	4	26	0	0	0	43	9
0	28	0	0	1	5	27	0	0	0	44	29
0	29	0	0	1	6	28	0	0	0	46	19
0	30	0	0	1	7	29	0	0	0	48	9
0	31	0	0	1	9	30	0	0	0	49	29
0	32	0	0	1	10	31	0	0	0	51	18
0	33	0	0	1	11	32	0	0	0	53	8
0	34	0	0	1	12	33	0	0	0	54	28
0	35	0	0	1	14	34	0	0	0	56	18
0	36	0	0	1	15	35	0	0	0	58	8
0	37	0	0	1	16	36	0	0	1	00	00

eotape. Since film length is usually listed or measured in "footage," I've supplied two tables for converting these footage measurements to videotape running time. Table 4.4 is for use with 16mm film. Table 4.5 is for use with 35mm film.

To use the tables, find the length of film footage in the left columns, then note the television time equivalent from the corresponding right-hand columns. For example, if you had a 16mm film clip that totaled 23 feet 26 frames, you would look up the foot and frame numbers in the left-hand columns of Table 4.4. Then, looking to the right-hand columns, you would note the following:

23 feet of 16mm film = 38 seconds and 9 frames of TV time.
26 frames of 16mm film = 1 second and 2 frames of TV time.

By adding the two conversions together, you get a total of 39 seconds and 11 frames of television time from 23 feet and 26 frames of 16mm film.

Preparing Film A/B Rolls for Television Transfer

When transferring film A/B (alternating scene) rolls to videotape, editors must remember the different frame rates of TV and film. Editors who attempt a straight rate transfer will discover that, after transfer and mixing, the videotape contains black frames at every film cut point, as well as picture instability at points where splices passed through the film gate.

Because film-to-tape transfer and subsequent mixing occur at a 24-to-30 (or four-fifths) frame ratio, it is necessary to include frame overlaps on all incoming and outgoing scenes that occur at A/B roll crossover points. An overlap of six to eight frames will make up for the frame rate difference, compensate for instability at film splice points and allow for slight cut point readjustments, if necessary, during the mixing process.

A/B rolls for film-to-tape transfer can be performed in three different ways, depending on which method suits the needs of a particular project.

• Transfer the entire amount of unedited footage (or just selected takes) onto two rolls of videotape simultaneously (double record). Projects to be edited and delivered on videotape with no film release could be transferred in this double record manner and then edited later. This method is also suited to projects that face tight delivery deadlines, since there is no delay while waiting for lab work to be finished.

• Untimed and unedited film scenes can also be recorded on two separate videotape rolls. This method might apply in cases where film workprints have been finished but no negative cutting has been done. The film editor keeps a foot/frame count log, which the video editor then uses for calculating film footage length to television time conversion. Using the log, the video editor locates the SMPTE time code number that corresponds to the "two pop" on the film leader. Then, to derive the correct edit frame number, the video editor adds the time conversion from Table 4.4 or 4.5 to the SMPTE time code number that corresponds to the film "two pop."

For example, with 35mm film, the editor might find that the "two pop" equals SMPTE time code 02:30:52:00 on the A-roll. In addition, if the film editor's footage log lists a cut point at 5 feet and 11 frames, the foot/frame conversion chart indicates that 5 feet equals 3 seconds and 10 frames and that 11 frames of film equals 14 TV frames. Adding the two durations gives us 3 seconds and 24 frames to be added to the reference reading of 02:30:52:00. Thus, the first cut point would occur at 02:30:55:24 on the A-roll transfer.

• The film editor may also transfer timed and edited (with overlap) A/B rolls, synchronized with a completed audio track. These A/B transfer rolls are then synchronized and mixed together during video editing by using the same footage/television time conversion method employed in the previous example. Some computer editing systems now feature functions that automatically convert foot/frame numbers to the correct time code reference.

Two final notes on A/B roll preparation. First, the footage needed for transitions (dissolves, wipes, etc.) should include the necessary eight-frame overlap counted from the *last* transition frame of the outgoing scene and the first frame of the incoming scene. Second, it is a common practice for film editors to use the midpoint of the transition for duration timing purposes, whereas videotape editors normally count from the first frame of the transition.

Contrast Ratios: TV versus Film

When transferring film to videotape, it's also important to remember that film and TV have very different contrast ranges (the range between the lightest and darkest picture elements). It is not uncommon to hear movie producers complain that films look more crisp

Table 4.5: Conversion of Film Footage to TV Time for 35mm Film

35mm Film		Television				35mm Film		Television			
Feet	Frames	Hours	Minutes	Seconds	Frames	Feet	Frames	Hours	Minutes	Seconds	Frames
0	1	0	0	0	1	39	0	0	0	26	0
0	2	0	0	0	2	40	0	0	0	26	20
0	3	0	0	0	4	41	0	0	0	27	10
0	4	0	0	0	5	42	0	0	0	28	0
0	5	0	0	0	6	43	0	0	0	28	20
0	6	0	0	0	7	44	0	0	0	29	10
0	7	0	0	0	9	45	0	0	0	30	0
0	8	0	0	0	10	46	0	0	0	30	20
0	9	0	0	0	11	47	0	0	0	31	10
0	10	0	0	0	12	48	0	0	0	32	0
0	11	0	0	0	14	49	0	0	0	32	20
0	12	0	0	0	15	50	0	0	0	33	10
0	13	0	0	0	16	51	0	0	0	34	0
0	14	0	0	0	17	52	0	0	0	34	20
0	15	0	0	0	19	53	0	0	0	35	10
1	0	0	0	0	20	54	0	0	0	36	0
2	0	0	0	1	10	55	0	0	0	36	20
3	0	0	0	2	0	56	0	0	0	37	10
4	0	0	0	2	20	57	0	0	0	38	0
5	0	0	0	3	10	58	0	0	0	38	20
6	0	0	0	4	0	59	0	0	0	39	10
7	0	0	0	4	20	60	0	0	0	40	0
8	0	0	0	5	10	61	0	0	0	40	20
9	0	0	0	6	0	62	0	0	0	41	10
10	0	0	0	6	20	63	0	0	0	42	0
11	0	0	0	7	10	64	0	0	0	42	20
12	0	0	0	8	0	65	0	0	0	43	10
13	0	0	0	8	20	66	0	0	0	44	0
14	0	0	0	9	10	67	0	0	0	44	20
15	0	0	0	10	0	68	0	0	0	45	10
16	0	0	0	10	20	69	0	0	0	46	0
17	0	0	0	11	10	70	0	0	0	46	20
18	0	0	0	12	0	71	0	0	0	47	10
19	0	0	0	12	20	72	0	0	0	48	0
20	0	0	0	13	10	73	0	0	0	48	20
21	0	0	0	14	0	74	0	0	0	49	10
22	0	0	0	14	20	75	0	0	0	50	0
23	0	0	0	15	10	76	0	0	0	50	20
24	0	0	0	16	0	77	0	0	0	51	10
25	0	0	0	16	20	78	0	0	0	52	0
26	0	0	0	17	10	79	0	0	0	52	20
27	0	0	0	18	0	80	0	0	0	53	10
28	0	0	0	18	20	81	0	0	0	54	0
29	0	0	0	19	10	82	0	0	0	54	20
30	0	0	0	20	0	83	0	0	0	55	10
31	0	0	0	20	20	84	0	0	0	56	0
32	0	0	0	21	10	85	0	0	0	56	20
33	0	0	0	22	0	86	0	0	0	57	10
34	0	0	0	22	20	87	0	0	0	58	0
35	0	0	0	23	10	88	0	0	0	58	20
36	0	0	0	24	0	89	0	0	0	59	10
37	37	0	0	24	20	90	0	0	1	0	0
38	0	0	0	25	10						

and vibrant during the screening than they do after transfer to videotape. What the producers mean is that the videotape version of the film doesn't seem to feature the same degree of separation between the shades of gray in the image—and they are right. The fact is, film is shot at a contrast range of 100:1 or better, while the TV contrast ratio falls somewhere between 15:1 and 30:1. This means that, with less separation between the lightest and darkest picture elements, the TV image simply can't reproduce all the subtle shades and shadows present in the film image. As a result, transferring film to videotape usually results in a loss of detail in low-light areas or a washing out of detail in high-light areas, depending upon how the telecine (film-to-tape) system is adjusted.

ASSEMBLING A POST-PRODUCTION TEAM

The division of responsibility among the members of an editing team varies from one project to the next. In some projects, a single editor may be responsible for making all post-production decisions and performing all editing functions. However, in most projects, the various roles and responsibilities are handled by a post-production team that can include the producer, the associate producer, the director or associate director, the script supervisor, the editor and the assistant editor. The personalities of the people filling these roles and the interpersonal skills of the editor become important factors in the success or failure of the post-production session.

To put it bluntly, editing drives many people up the wall. The intense hours spent in an editing bay under the pressure of project deadlines can strain the best of friendships and quickly transform relationships that are less than ideal into full-fledged animosity.

To make post-production go as smoothly as possible, it is important that editors and producers understand the roles various team members should play and that they find the right people to fill those roles. I prefer to adopt the philosophy that "we're all in this together" and try to make the team a cohesive unit working toward the same objectives. As I've told many clients, "We're going to get from point A to point B regardless. What matters is *how* we get there."

For producers and directors entering into video post-production for the first time, I strongly recommend that they sit down and talk with the editor—preferably before shooting the project. That way, they can ask any questions pertaining to the post-production process and find out about any special situations that need to be included in the production shoot (chroma keys,

special effects sequences, two-track or stereo audio capabilities, etc.). I used to recommend that directors bring someone experienced in post-production along with them to make them feel more comfortable. But they would invariably bring along their film editor, who often knew less about videotape post-production than they did.

The Producer

The producer has the ultimate responsibility for a project. However, many producers normally don't attend the editing sessions. Instead, they order cassettes of the various editing stages and view them at their convenience. This is a more efficient process for both the producer and the editor, as anyone who has struggled through editing amidst a chorus of ringing phones can attest.

The Associate Producer

The associate producer is responsible for budgeting and scheduling the project, for making sure it all comes together as planned and for patching it back together when it comes apart at the seams.

A good associate producer knows from experience the best way to get what is needed, where to get it and how to deal for it.

The Director and the Associate Director

The director is responsible for the visual continuity and the performances in the project. Most commercial and industrial directors take an avid interest in the video post-production process, since they have been deeply involved in the project from the start.

However, on many TV series projects, the director does not sit through the editing process. Production schedule conflicts usually require the associate director to represent the director throughout post-production. Initially, the director, the associate director, the editor and the producer will meet and prescreen the material to select takes. Then, after a first rough cut is completed, the director will review it.

The Script Supervisor

The script supervisor is responsible for all the production notes to be used during editing. These include scene and take time code, reel locations, continuity and technical notes, director's notes, camera shot descriptions, and segment lengths. In addition to being familiar with the location of valuable performance pickups, music track layovers and important isolation shots, script supervisors often function as the producer's private secretary, fielding and channeling all incoming calls and relaying messages.

The Editor

The editor's main job is to please the clients. This requires the editor to possess creative and technical skills, an even temper and the ability to function as both psychologist and teacher. On those frequent occasions where an editor and clients are working together for the first time, there is a natural "breaking in" period. During this period, as the editor is getting the equipment turning, he must also analyze the clients and the situation. Who's in charge? Are they in agreement? Do they have all the materials they need to finish the job?

Just as important, he must become familiar with the jargon the client uses for editing instructions. For instance, when the director says to "go ahead" on a playback tape, does he mean to go ahead into the material further or to go back toward the beginning (or "head") of the reel. Believe me, I've heard the term used both ways.

On top of his other responsibilities, a staff editor is responsible for keeping track of which equipment is used and for how long. This is the information the editing facility uses for billing the clients, so inaccurate record keeping results in needless misunderstanding and a probable loss of revenue for the facility.

The editor should be familiar enough with editing equipment that he doesn't have to concentrate on the operating procedures during the edit session. However, with new, sophisticated equipment coming on the market so quickly, it's hard to keep up. I remember coming to work on a Monday morning not too long ago to find myself scheduled to edit several commercial spots with a digital effects system that wasn't even installed in the editing bay on the previous Friday. Consequently, I spent the half hour before the session madly scanning the instruction manual. Later, I gave the client a reduction on the editing charges, since the commercial session took longer than it should have. In other words, I was honest with the client.

Honesty is probably the single most important factor involved in a successful editor-client relationship. If an editor can combine honesty and competence, he should have no trouble establishing cordial, long-term relationships with clients. In fact, some clients come to rely exclusively on one editor, making the editor feel that he is indispensable in their eyes.

But indispensability is a myth that only a fledgling editor or egotist would fall for. This was graphically demonstrated to me years ago in a colleague's almost fatal experience. After several extremely long days of editing, including two all-night sessions, he was near the point of exhaustion. With the producers fervently pleading with him to stay—"No one else can handle it as well!"—he reluctantly agreed, only to collapse minutes later from complete exhaustion. As the ambulance attendants wheeled him out, the producers were already going full speed ahead with a different editor.

The Editor's Assistant

Online editing remains the most technically exacting phase of videotape post-production. The editor's assistant must be aware of all the technical aspects of online editing—in other words, all the technical factors involved in the process of assembling a final product on videotape.

The assistant's duties include much more than just loading a VTR and knowing how to set it up for playback. Since the assistant is stationed in the equipment room with the videotape machines, it is his responsiblity both to load the proper playback reels for the editor and to check for technical details such as color framing, horizontal sync timing of the playback VTRs, and video and chroma levels. The assistant should also be well versed in operating any other electronic equipment found in the editing bay, including color or matte cameras, slow-motion disk machines, and mag track or audio tape machines.

An alert and competent assistant can make a great difference in a post-production session. By keeping track of numerous technical details, the assistant leaves the editor free to concentrate on working with the project representatives and supervising the overall editing effort. The editor and the assistant must work as a team, and, as always, good teamwork depends on developing a cohesiveness based on professional expertise and mutual trust.

CONCLUSION

Let me conclude this chapter as I began it, by emphasizing the importance of careful preparation in the post-production process. By adhering to the planning principles described on the preceding pages, editors and directors can avoid many of the costly pitfalls that await those who blindly plunge ahead. In my experience, time spent on preparing budgets, selecting videotape stock, arranging film-to-tape transfers and organizing a post-production team is paid back in time and money saved during the actual editing sessions.

Above all, editors must establish channels of communication with the producer and with representatives of the production company, so everyone understands the goals of the editing session and so there are no surprises when the final product is delivered.

5 The Offline Editing Process

In its most general sense, the term "offline editing" includes any editorial activity that does not involve using the original production footage and creating a finished master tape. Under this broad definition, offline editing encompasses viewing production footage, selecting takes, taking production notes, hand logging edits, and editing rough- and fine-cut workprints.

However, "offline editing" is usually used in a much more narrow sense, to refer only to the use of lower-cost editing systems and duplicate tapes ("window dupes") of production footage to assemble workprints and an edit decision list. The 3/4-inch or 1/2-inch workprints are then sent to producers, network executives or company officials for final approval before committing the project to online editing, when the original production tapes and a revised (or "cleaned") edit decision list will be used.

The tremendous advantages of offline editing were demonstrated to me when I edited three dramatic programs for the "Wide World of Entertainment" series broadcast on ABC. The three programs each contained an equal number of edits and approximately the same amount of production footage. In addition, the programs were scheduled to run for the same length of time with an identical number of commercial breaks.

The first program was edited entirely online, according to the normal procedures then in place, with the new (at the time) CMX 300 computer edit system. The final billable online edit time came to 26 hours—a great improvement over the 40 to 50 hours of editing time

required with earlier, non-time-code online systems.

For the second program, we added an offline editing session. The offline editing was performed on two 1-inch VTRs, and the edit point numbers were logged by hand for subsequent entry on the CMX 300 online system. By adding the offline session, we were able to cut the online edit time in half—to 13 hours.

For the third program, the offline editing was performed on the CMX 400 1-inch computer offline editing system, which had just been introduced. Editing on the CMX 400 allowed us to produce a frame-accurate workprint, a paper punch computer tape and an edit decision list readout. The punch tape list was then fed into the computer memory of the compatible CMX 300 system for subsequent automatic editing, using sequential A mode assembly. With these changes, the third program required only eight hours of online editing time.

Even counting the added offline editing time (which is billed at a considerably lower rate per hour), the third program cost far less to edit than the first two. In addition, for the third program the director was able to make creative changes after viewing the frame-accurate workprint. In the earlier editing sessions, in which there was no offline workprint, creative changes could only be made after the actual online session. This generally resulted in a loss of a generation (and a degradation of video quality) on the final edited master.

In this chapter we'll look at the factors that contribute to successful offline editing, including the decision of which type of editing system to use and standards for preparing workprint reels. The major emphasis of the chapter is the preparation and management of edit decision lists, whether hand-logged or computerized.

TYPES OF OFFLINE EDITING

The three main types of offline editing—control-track, automatic time-code and computerized time-code—were described in Chapter 3. In this section we'll review their usefulness for different offline editing situations.

Control-Track Offline Editing

If you consider only per-hour equipment costs, control-track editing is by far the least expensive form of offline editing. In 1987, the rental rate for a simple, two-VCR, 1/2-inch or 3/4-inch, cuts-only control-track system, similar to the one shown in Figure 5.1, was approximately $750 per week. Pretty reasonable, considering that this allows an editor to use the system 24 hours a day for seven full days. But price is not

the whole story. As discussed in Chapter 3, control-track edit systems require the editor to cue each source tape manually to the proper edit points and to log all edit decisions by hand.

In a facility where control-track videocassette editing systems are used for both offline and online editing, these limitations do not prove to be too serious. However, for those projects that will move on to time-code-based online editing, control-track offline editing can end up costing more money than it saves. In particular, there is the cost of the extra time required to enter the hand-logged edit numbers into the computer's memory, as well as the risk of error this introduces into the editing procedures. Generally, the more complex the editing project, the less likely it is that using control-track equipment for offline editing will pay off in actual cost savings.

Automatic Time-Code Offline Editing

Automatic offline editing using time-code equipment, like that shown in Figure 5.2, is usually a faster and more efficient process than control-track offline

Figure 5.1: Control-track editing using the JVC VE-93 editing controller and JVC CR-850U editing VCR. Photo courtesy of JVC Co. of America.

Figure 5.2: The Convergence ECS-104 editing controller used in automatic time-code editing. Photo courtesy of Convergence Corp.

editing. Depending on the sophistication of the equipment, automatic time-code offline editing can offer the following advantages:

• Entering edit points in time code, so several VTR playbacks can be automatically cued to the correct locations;

• Controlling all VTR and edit functions from a single control panel;

• Performing split edits in a single edit function;

• Automatic printing of edit decision lists; and

• Automatic preparation of edit decision lists in computer-readable form (punch tape or floppy disks).

In short, compared to control-track systems, time-code editing systems allow editors to perform more offline chores automatically, so the editor can spend more time concentrating on the aesthetics of the project. In addition, time-code systems make it much easier for editors to prepare accurate, computer-readable edit decision lists.

Computerized Offline Editing

For the most part, the major computerized systems used in offline editing, such as the one shown in Figure 5.3, are identical to their online counterparts. The only real differences are that offline systems are capable of fewer video switcher effects and they use small-format VTRs. In fact, for projects originally produced on 3/4-inch tape, 3/4-inch offline editing systems often double as online systems.

For editors and directors involved in projects that will employ computerized offline systems, I strongly recommend thorough screening of the production material, including the logging of rough edit points, before the actual offline session. The screening room should be equipped with a playback VTR compatible with the format (3/4-inch, 1/2-inch, etc.) of the work-print tapes, preferably a VTR that features a controllable still-frame function, and a color television monitor. In fact, on most editing projects, I try to

Figure 5.3: A multi-cassette computerized editing bay. Photo courtesy of Vidtronics, Inc.

have two VTR/monitor playback systems set up, to make it easier to compare multiple takes of the same scene recorded on different reels. Nothing is as frustrating as trying to remember how a particular take looked while changing reels and cueing up a second tape.

When preparing to edit a project offline, you should ask first of all how elaborate a system you really need. If your project is recorded on just four or five reels of tape and requires only a few basic transitions, why use a computerized system with four playback VTRs and a video switcher equipped with elaborate special effects capabilities?

Today, with many network shows being shot using four isolated cameras (cameras whose video signals are recorded on individual VTRs), some producers seem to believe that they must edit with four playback

machines. Apparently, they feel that editing will be quicker and more efficient with the four camera reels loaded into individual playback machines. The impractical implications of this belief become apparent when you realize that a producer probably also has available a switchfed reel recorded on a line VTR plus several reels recorded at dress rehearsal. Should individual VTRs be used for each of these reels, too, resulting in an editing setup with 10 or more playback VTRs?

In my experience, three playback machines will almost always do the job for four-camera productions. For corporate or educational productions, two playback VTRs will usually suffice. On those network shows that use line switchfed reels plus isolated camera reels, I usually put the switchfed reel in one playback VTR and whatever isolated camera reels I need on second and third playback VTRs. On those projects in which there are no switchfed reels, I usually keep the isolated reels from cameras one and four loaded on playback VTRs at all times, since these cameras are generally used for most cross-two and close-up shots. This leaves the third playback VTR for the reels from cameras two and three—the cameras that are normally used for wider shots. I alternate between the two reels, loading the reel needed at the time into the third VTR.

PREPARATION OF WORKPRINT REELS

Workprint reels (or window dupes), which have time code numbers inserted in the picture, are generally recorded on 3/4-inch videocassette stock, although 1/2-inch Beta and VHS formats are beginning to find favor at some production and post-production companies. For most in-studio productions, workprint reels are recorded simultaneously with the master production reels. This eliminates the need for recording costly duplicates (or "dupe-downs") at the post-production facility. However, on most remote location shoots, it does not make sense to bring along the additional VTR equipment required to record simultaneous workprints. As a result, producers and directors who are planning remote location projects should also plan on spending the extra money required to record cassette dupedowns at the post-production facility.

One advantage of recording dupedowns in post-production is that problems with production reels are discovered before online assembly begins. For example, I was involved in editing an in-studio production for which the workprint reels were recorded simultaneously with the production reels. While using the workprint reels during offline editing, we noted a particularly effective swish-pan transition that we decided to use several times in the final online session. Unfortunately, once we began online assembly with the original production reels, the swish-pan was nowhere to be found. Apparently, the videotape operator had inadvertently stopped the production reel VTR before the swish-pan was recorded.

Criteria for Preparing Workprint Reels

Whether you choose the simultaneous or the dupedown method for preparing workprint reels, the reels should be recorded in accordance with the following criteria of the accepted TV standard format for offline workprints.

Playback Material

The video track should consist of program video with character-generated time code numbers super-imposed in the picture. The numbers should be large enough to read comfortably, but small enough to be unobtrusive.

Audio track 1 should consist of program audio with test tone at the head of tape, recorded at 0 volume units (VU), as measured on the cassette VTR's audio meter.

Audio track 2 should be the same as audio track 1 for control-track or BVU cassette recordings. For time-coded workprints, SMPTE time code should be recorded at a +2VU level, as measured on the cassette VTR's audio meter during playback.

When applicable, the address track should consist of SMPTE time code recorded at a +2VU level.

Record Edit Master Stock

The video track should contain video black with color burst and sync signals.

Audio track 1 should contain no audio.

Audio track 2 should be the same as audio track 1 for control-track or BVU cassette recordings. For time-coded reels, SMPTE time code should be recorded at a +2VU level, as measured on the cassette VTR's audio meter during playback.

Workprints: Six Important Facts To Check

1. Make sure that the time code being recorded on audio track 2 or the address track is the same code that is triggering the character generator numbers in the picture. Otherwise, confusion arises as to which code is correct. Offline editing may also be done without time code numbers in the picture, but this makes for slower and less efficient offline sessions. One method I use for clients who prefer not to have code numbers in the picture during screening sessions is to put the numbers outside of the television safe action area. That way, the numbers will not appear on the monitor screen—unless the monitor is set in the "underscan" mode (see Figure 5.4).

2. Be sure that the offline videocassette is a duplicate of the program master videotape. As I mentioned earlier, there have been cases in which shots on the offline cassette did not match the shots on the master videotape used in online editing. As you can imagine, this causes considerable problems during the online

Figure 5.4:   A workprint reel recorded with the time code numbers outside the safe action area. Photo by Darrell R. Anderson.

session. The time code recorded on the master videotape and the offline videocassette must also be identical. This sounds obvious, but mismatched time code continues to be a frequent problem in video editing.

3. Check that the time code generator is synchronized to the input video. Unsynchronized time code will drift, making it impossible for a computerized editing system to perform the edits.

4. Remember that the time code and the video signal sent to the offline cassette VTR should not simply be fed through the production VTR and recorded directly on the workprint reel—this would normally result in the time code being out of phase with the video signal, due to the time delay introduced by the production VTR's electronic circuitry. Remember, too, that out-of-phase time code means that the sync word of the code will be out of line with the vertical sync pulse of the video signal (see Chapter 2).

5. Keep in mind that, to use address-track SMPTE time code, the editing system must feature BVU cassette VTRs that offer an address-track function.

6. Finally, be sure to use videocassette stock that features still-frame capability. Lesser quality cassette stock tends to crease and tear under the stress of editing.

LOGGING OFFLINE EDIT DECISION LISTS BY HAND

On most small, offline control-track edit systems, editors must hand log the edit points. When this is the case, the editor needs to be concerned only with the edit points on the original program material. Most editors devise their own individualized method of logging edits. The only universal requirements are that the edit decision list (EDL) is organized in some logical order and that it can be readily interpreted for quick and efficient online assembly.

The basic information needed on an EDL includes:

1. Edit numbers for each shot, numbered consecutively.

2. A reel number for each shot.

3. The edit mode (audio-only, video-only, or audio *and* video), plus the number of the audio track on which audio edits will be recorded.

4. The transition type (cut, dissolve, wipe, key, etc.).

5. The time code of the edit start point.

6. The duration of the edit (for timing purposes).

1 #	2 Reel	3 Mode	4 Transition	5 Start	6 Duration	7 Shot Description
01	33	A1V	C	02:00:10:00	5:00	X 2 shot
02	34	V	C	03:10:00:00	10:00	Close up single
03	33	A1	C	02:10:15:12	6:00	Narration
04	34	A1	C	03:10:06:00	4:00	Sync dialogue
05	32	A1V	D030	06:25:52:00	18:00	3 shot med
06	32	A1V	C	06:27:35:20	9:00	Wide shot/location snd efx
07	31	A2	C	01:12:25:24	9:00	Narration

Figure 5.5: A sample hand-logged edit decision list.

7. A description of the shot (for reference).

Figure 5.5 is a sample edit decision list. As you can see, the list contains all seven of the basic information items needed for online assembly. The video-only edits and the corresponding audio that will be recorded during the insert edit are bracketed together. The same would be done for audio-only edits. Thus, edits 3 and 4 (audio) add up to the total length of the video insert in edit 2.

Looking at edit 5, you can see that it includes a dissolve of 30 frames duration starting at 06:25:52:00. (However, some editors prefer to note the *midpoint* of the dissolve, as is done in film editing; be sure it is clear which method is being used.) The outgoing video needed for the dissolve will come from edit 2 (03:10:10:00), while the outgoing audio will come from edit 4 (03:10:10:00). In this case, it happens to be a relatively simple single-source audio/video dissolve. Edits 6 and 7 require audio narration, recorded on audio track 2, simultaneously with both video and location sound effects, recorded on audio track 1.

Of course, Figure 5.5 is only one possible EDL format. Most editors using small offline systems add their own personal touches, or they adopt a standard list form used by a particular editing facility. In the final analysis, when you find a type of list that works for you, by all means use it.

COMPUTERIZED EDIT DECISION LISTS

Although SMPTE has formed a committee to work on establishing a standardized industry-wide format for computerized edit decision lists, no single standard exists today. Currently, the most widely used EDL format is the ASCII (pronounced ''as-key''; American Standard Code for Information Interchange) format,

more commonly referred to as the CMX format. As the various edit decisions in the sample EDL in Figure 5.6 indicate, the basic ASCII format consists of ten discrete areas of information. These areas of information, or data fields, are described below.

1. Field 1 contains the edit number. This indicates the numerical order in which the edits in the list will be performed.

2. Field 2 indicates the source of the edit. This may be a reel number containing up to six alphanumeric characters or abbreviations such as ''AX'' for auxiliary, ''BL'' for black or ''ISO'' for isolated camera.

3. Field 3 indicates the edit mode. This includes designations for audio/video, audio-only, dual track audio, and video-only edits.

4. Field 4 designates the edit transition type. This can include any of the following letter/ number combinations:

C   Cut

D   Dissolve transition. This is a two-line edit, with the first line indicating the frame reference for the outgoing scene.

W***   Wipe-transition. This includes a three-digit number signifying which wipe configuration the video switcher should perform.

KB   Key transition. The B indicates that this is the first line of a two-line edit and is the background source for the edit.

(1)	(2)	(3)	(4)	(5)	(6)	(7)	(8)	(9)	(10)
001	ISO A	AV	C		01:45:45:00	01:45:50:00	01:00:00:00	01:00:05:00	
002	ISO A	AV	C		01:45:50:00	01:45:50:00	01:00:05:00	01:00:05:00	
002	LINE	AV	D	090	01:45:50:00	01:46:00:00	01:00:05:00	01:00:14:28	
003	LINE	AV	C		01:46:00:00	01:46:00:00	01:00:14:28	01:00:14:28	
003	ISO B	AV	W101	060	01:53:31:00	01:53:46:00	01:00:14:28	01:00:29:28	
004	ISO B	AV	K B		01:53:46:00	01:53:56:00	01:00:29:28	01:00:39:28	
004	TITLE	AV	K	015	00:00:00:00	00:00:05:00	01:00:29:28	01:00:34:28	
005	ISO B	AV	K B	(F)	01:53:56:00	01:54:06:02	01:00:39:28	01:00:49:28	
005	TITLE	AV	K O	015	00:00:10:00	00:00:10:00	01:00:39:28	01:00:39:28	
006	ISO A	V	C		01:46:00:00	01:46:05:00	01:00:49:28	01:00:54:28	
007	ISO B	A2	C		01:54:06:02	01:54:11:02	01:00:49:28	01:00:54:28	

Figure 5.6: An edit decision list showing the ten ASCII data fields.

K   Key-in transition. K by itself indicates that this is the second line of a two-line edit and that it is the foreground source for the edit.

KO   Key-out transition. The O indicates that this is the second line of a two-line edit and that it is the foreground source for the edit. KO differs from the K in that the foreground will be on when the edit begins, and will then fade out after the prescribed duration.

5. Field 5 can either be blank (as would be the case for a cut), or it can contain three digits of information, ranging from 000 to 255. These digits designate the length, in frames, of the transition. Field 5 may also contain the letter ''F,'' signifying a fade to or from black of both the foreground and background image during a key transition edit.

6. Field 6 indicates the start time of the playback VTR, designated in SMPTE time code.

7. Field 7 indicates the stop time of the playback VTR, designated in SMPTE time code.

8. Field 8 indicates the start time of the record VTR, designated in SMPTE time code.

9. Field 9 may indicate either the stop time of the record VTR or the duration of the edit. This data field is really just a convenience and has no real relation to the definition of the edit.

10. Field 10 contains the carriage-return and line-feed information necessary for the computer to begin reading the next edit in the list. However, this information is not actually printed in the edit decision list readout.

Many computer edit decision lists can and do contain additional amounts of data, but this basic ASCII format generally remains intact. In the future, the list format may expand, particularly as industry needs develop. For instance, now that multiple-audio-track VTRs are becoming common in video post-production, there is much discussion about including a field for designating the audio track.

Even now, most edit system manufacturers are devising their own designations for edit decisions and data fields. Figure 5.7 includes one example of this manufacturer's format—an edit decision list for the Convergence Corp. ECS-104 edit system. As the figure shows, Convergence has designated data field 3 for identifying audio tracks by numerical designation. Edit 1, for instance, identifies the edit as recording both video and audio track one, whereas edit 16 will record audio only, on both tracks 1 and 2. The decision lists of other manufacturers include comparable variations.

In the future, EDL designations will undoubtedly include data to set up and activate digital video effects devices, to identify special features of video and audio switching equipment and to control variable speed motion features. This information would, of course, be entered during the offline editing process.

```
* PROCESSED EVENT LIST, NON-DROP-FRAME FORMAT

* CONVERGENCE 104 AUDIO FORMAT
* LIST HAS NOT BEEN CLEANED
   (1)    (2)   (3)  (4)  (5)      (6)            (7)            (8)           (9)
  001 ABC123 V1   C          01:42:25:05 01:42:30:05  01:03:39:14 01:03:44:14

  002 2      V1   C          01:42:00:09 01:42:05:09  01:03:44:14 01:03:49:14

  003 3      V1   C          02:05:25:18 02:05:30:18  01:03:49:14 01:03:54:14

  004 ABC123 V2   C          01:42:30:05 01:42:35:05  01:03:54:14 01:03:59:14

  005 2      V2   C          01:42:05:09 01:42:10:09  01:03:59:14 01:04:04:14

  006 3      V2   C          02:05:30:18 02:05:35:18  01:04:04:14 01:04:09:14

  007 ABC123 V12  C          01:42:35:05 01:42:40:05  01:04:09:14 01:04:14:14

  008 2      V12  C          01:42:10:09 01:42:15:09  01:04:14:14 01:04:19:14

  009 3      V12  C          02:05:35:18 02:05:40:18  01:04:19:14 01:04:24:14

  010 ABC123 A1   C          01:42:40:05 01:42:45:05  01:04:24:14 01:04:29:14

  011 2      A1   C          01:42:15:09 01:42:20:09  01:04:29:14 01:04:34:14

  012 3      A1   C          02:05:40:18 02:05:45:18  01:04:34:14 01:04:39:14

  013 ABC123 A2   C          01:42:45:05 01:42:50:05  01:04:39:14 01:04:44:14

  014 2      A2   C          01:42:20:09 01:42:25:09  01:04:44:14 01:04:49:14

  015 3      A2   C          02:05:45:18 02:05:50:18  01:04:49:14 01:04:54:14

  016 ABC123 A12  C          01:42:50:05 01:42:55:05  01:04:54:14 01:04:59:14

  017 2      A12  C          01:42:25:09 01:42:30:09  01:04:59:14 01:05:04:14

  018 3      A12  C          02:05:50:18 02:05:55:18  01:05:04:14 01:05:09:14

  019 ABC123 V    C          01:42:55:05 01:43:00:07  01:05:09:14 01:05:14:16
```

Figure 5.7: A sample dual audio format edit decision list for the Convergence ECS-104 edit system.

COMPUTER CARDS, PUNCH TAPE AND FLOPPY DISKS

Edit decision lists can be stored in a number of computer-readable forms. In this section, I describe the advantages and limitations of the three most common forms: computer cards, punch tape and floppy disks.

Computer Card Conversion

The punch card system was actually an intermediate step in the progression from hand logging edit deci-

sions to fully computerized list output capabilities. The punch card system was initiated by the late Hal Collins, as a means of updating his 1/2-inch offline editing system (see Chapter 1). The process involved hand logging each edit decision, as was the previous practice, then transferring that data onto computer cards via a key-punch machine.

Two Hollywood post-production facilities, Vidtronics, Inc. and Consolidated Film Industries (CFI), developed computer software programs that would convert the punch card data into a decision list compatible with CMX computerized online editing systems. Four D Productions (the producers of "Barney

Miller'') also used this computer card system for several seasons, until their switch to fully computerized offline editing on the CMX 340X and Fernseh Mach One systems. Since most other production and post-production companies have followed suit, I know of no major editing facility that currently uses the punch card system.

Punch Tape

Before the introduction of magnetic floppy disks, all final edit decision lists used for automatic assembly were compiled on paper punch tape, with a hardcopy readout printed on paper by a line printer. (See Figure 5.8.) Although punch tape is still being used as the primary list format at some facilities, I personally prefer to use punch tape only as a backup for the floppy disk, to insure against the necessity of retyping every edit in the event of a disk damage or computer failure.

The punch tape used in video editing is a standard 1-inch wide thin paper tape. It comes in either 1000-foot rolls or in boxes of 1000 feet of fanfold tape, depending on which type your reader/punch mechanism is designed to use.

Each byte (eight bits of data) is punched into one frame on the tape. A frame consists, then, of eight data positions arranged in a line perpendicular to the length of the tape (see Figure 5.9). A hole punched in the tape equals ''1,'' while no hole equals ''0.'' (See the discussion of binary code in Chapter 2.) Paper tape is designated as an eight-channel format, with each channel comprised of one data position in each frame.

When loading a punch mechanism with fresh paper tape, take extra care that the tape is moving through the mechanism's guides properly. The unwelcome consequences of misaligned tape include the possibility of a jammed feeder or, even more disastrous, feed holes that were not punched properly. This last problem results in a punch tape that will not read properly when fed into the tape reader during online assembly.

I have also found, to my dismay, that punch tape makes a marvelous sponge when coffee is spilled on it. However, if it is accidentally torn, paper tape can be spliced together with little problem by using any standard paper tape splicing block.

Floppy Disk Systems

Floppy disk systems are the latest innovation for storing and recalling edit decision lists. At many editing facilities, disk configurations are rapidly replacing the paper tape systems used through the late 1970s. The disks are reusable, they are capable of faster data input/output, and they have a larger practical edit list capability—with paper tape systems, long EDLs won't fit into the tape reader device.

The disks used in computerized editing systems come in two different diameters: 8-inch and 5 1/4-inch. During offline editing, be sure to use a disk that will fit the disk drive unit used on the online editing system. The disk itself is an oxide-coated nylon platter that is permanently mounted inside a protective paper envelope. When the disk is inserted into the disk drive, it rotates at 360 rpm so that information can be written to or read from it (see Figure 5.10).

Editing systems record on floppy disks in one of two formats: single-density or double-density. Both use the same type of disk; the difference is the data encoding process used to format and record the data bytes on the disk. The single-density disk can store up to 256K (kilobytes) of data. In practical terms, this translates into the ability to store an edit decision list of slightly more than 4000 edits. The double-density disk can store up to 512K of data— or an edit decision list of slightly more than 8000 edits.

Originally, single-density formatting was preferred by most editors, simply because it was more reliable. After all, double density formatting crams twice as much data in the same amount of disk space, requiring the read/write mechanism of the disk drive unit to be

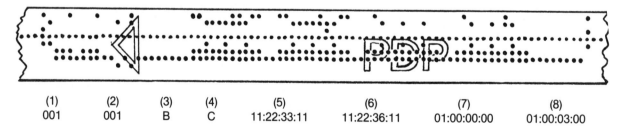

(1)	(2)	(3)	(4)	(5)	(6)	(7)	(8)
001	001	B	C	11:22:33:11	11:22:36:11	01:00:00:00	01:00:03:00

Figure 5.8: Punch tape with corresponding hardcopy readout.

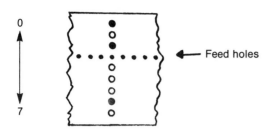

● Hole unpunched (value of 0)

○ Hole punched (value of 1)

Figure 5.9: The eight-channel punch tape format, showing a frame, or byte, of data.

much more sensitive and accurate. However, as disks and disk-drive units have improved, double-density formatting has become more reliable and, as a result, more popular. Currently, some editing systems use only single-density disks, some systems use only double-density disks, and some are equipped to handle both.

MANAGING YOUR EDIT DECISION LISTS

During the offline editing process, edits are continually changed and recorded over. Consequently, your final offline edit decision list will contain a number of edits that have become "obsolete." Editors call this a "dirty" list. Previously, it was necessary to spend several hours "cleaning" dirty lists. Overlapping edits had to be trimmed back and undesired edits deleted. Under the pressure of deadlines, many editors became hopelessly lost. Often, they would simply assemble every edit, good and bad, and hope that the show came out correctly in the end.

But that was before. Today, most modern edit systems with memory capability feature something called "list management modes of operation." With this function, the computer allows the editor to recall any event stored on the edit decision list. Once they are recalled, the edits can be changed, deleted or shifted to different locations on the list. In addition, the editor can change the duration of an event, and the computer will automatically update all of the following edit points and record times. (This is called "rippling" the list.) Groups of edits can also be shifted to any other location on the list. For example, an editor might use this function to change an entire song or scene from one location in the show to another.

Although the list management mode made it far easier for editors to modify and clean up an edit decision list, it still required editors to make changes one edit at a time (with the exception of group edit shifts). This limitation was resolved by the next development in edit list management—the 409 list management system.

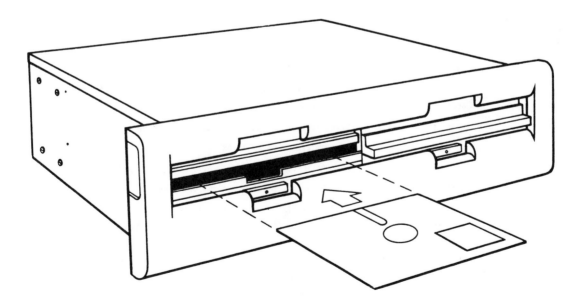

Figure 5.10: Inserting a floppy disk into the disk drive. Courtesy Data Systems Design.

Edit List Cleaning with the 409 Program

In 1976, working in cooperation with Vidtronics, Inc., computer programmer Dave Bargen developed a piece of software that has come to be known as the 409 program. When loaded into a standard editing computer equipped with at least 16K of memory, the 409 program allows editors to clean edit lists in a fraction of the time required with other methods of list management.

The 409 program works like this: A dirty edit decision list, like the one shown in Figure 5.11, is entered into the editing computer's memory. The dirty list can be loaded from either punch tape or a floppy disk—just as long as it was generated on an ASCII-compatible system. In fact, the 409 program can clean several edit decision lists simultaneously, as long as the lists are entered into the computer in record-time sequential order.

After the dirty list has been loaded into the computer, the editor merely activates the 409 program by pressing the proper key indicated on the display terminal menu as the Automatic Cleaning Mode. Several individual modes of list cleaning are also featured, such as:

• Clean only—eliminates out-edit point overlap.

• Join only—combines separate audio and video edits that are in sync and have the same reel number.

• Video only—eliminates all audio edits in the EDL.

• Audio 1 only—eliminates all video and audio 2 edits in the EDL.

• Audio 2 only—eliminates all video and audio 1 edits in the EDL.

• Trim input times—trims the in and out edit times on the source material and/or the record reel to compensate for any system EDLs that have built-in time code error offsets.

• Trim output times—trims the source and record out edit times to compensate for any inadvertent gaps appearing in the EDL.

• Time code format—409 has a drop frame, non-drop frame and SMPTE mode. The SMPTE mode will clean either all drop frame EDLs, all non-drop frame EDLs, or EDLs containing a mixture of both.

The end product of the 409 program is a clean edit decision list—a list that can be used for online assembly in either the sequential A mode or the checkerboard B mode; both were described in Chapter 3. Figure 5.12 shows the cleaned version of the EDL in Figure 5.11. Notice that the last column indicates the duration

```
0001 20     A12V C       19:54:19;23  19:54:36;02  01:01:34;14  01:01:50;23

0002 901    A12V C       01:01:22;14  01:02:07;00  01:01:34;14  01:02:19;00

0003 20     V    C       20:01:05;17  20:01:25;16  01:01:34;14  01:01:54;13

0004 21     A2   C       20:01:43;06  20:01:45;17  01:02:55;06  01:02:57;17

0005 21     A12  C       20:01:42;10  20:01:43;03  01:02:57;13  01:02:58;06

0006 21     V    C       20:01:40;03  20:01:44;04  01:02:53;27  01:02:57;28

0008 21     A2   C       19:55:01;07  19:55:03;04  01:02:55;06  01:02:57;03

0007 20     A1V  C       20:02:00;05  20:02:04;27  01:02:58;04  01:03:02;28

0009 901    A12V C       01:02:28;16  01:03:31;28  01:02:39;13  01:03:42;25

0010 901    A12V C       01:03:31;25  01:03:31;25  01:03:42;22  01:03:42;22
0010 200    A12V W001 060 20:11:45;20  20:11:51;26  01:03:42;22  01:03:48;28
```

Figure 5.11: A "dirty" edit decision list (ASCII format).

```
001 901 A12   C      01:01:22;14 01:02:07;00   01:01:34;14 01:02:19;00   00:00:44;14

002 20  V     C      20:01:05;17 20:01:25;16   01:01:34;14 01:01:54;13   00:00:19;29

003 901 V     C      01:01:42;13 01:02:07;00   01:01:54;13 01:02:19;00   00:00:24;15

004 901 A12V  C      01:02:28;16 01:03:31;25   01:02:39;13 01:03:42;22   00:01:03;09
004 200 A12V  W001 060 20:11:45;20 20:11:51;26   01:03:42;22 01:03:48;28   00:00:06;06
```

Figure 5.12: The edit decision list from Figure 5.11 after cleaning with the 409 program.

of each event; this is for reference only and is not part of the active edit decision list.

The 409 program is both easy to operate and extremely versatile. Over the years, on the basis of extensive practical experience, the program has been preset to accommodate the most frequently used operating modes. For instance, the 409 program assumes that:

- The EDL is recorded on a floppy disk.

- The time code mode will be SMPTE.

- Automatic full cleanup is preferred.

- The cleaned list will be returned to the floppy disk.

Of course, the preset (or default) conditions can easily be changed by the editor to adapt the 409 program to the conditions required by any given editing situation. Other features of the 409 program include the ability to:

- Print out a hard copy of the EDL.

- Punch the EDL out on paper tape.

- Enter new EDL titles on cleaned lists.

- Retain notes in the EDL or delete notes in the EDL.

- Provide a reel summary to indicate the amount of edits from each reel used in an assembly, as well as a listing of each transition edit and which reels are involved (see Figure 5.13).

A separate translator program (XEDL) is provided by The Grass Valley Group (now the owners of 409)

to customers buying the 409 program. This allows them to convert between non-compatible EDLs and The Grass Valley Group (GVG) format, which is the only outputted format of the 409.

The Trace Program

The trace program is another effective list management tool developed by Dave Bargen. With the trace program, an editor can make several generations of changes to a project, enter the information from the various generations into the computer and end up with a final edit decision list that contains time-code references matched to the original source reels.

For example, assume you're an editor preparing for the second cut of a large project. You certainly aren't going to go back and remake every edit on the list, and you can't splice the videotape to open up a scene. Instead, you should use the rough-cut edited master as a playback reel, assign it a reel number and use it to edit a second master reel.

In other words, to create a second-cut master, you would edit down a generation. Thus you would be using both the original playback material and the first-cut master to put together the second-cut master. For a third cut, follow the same procedure. As you can probably surmise, the third-cut reel would contain material from the original master, the first-cut master and the second-cut master. The key to this "editing down" process is *to assign each cut master a reel number—a number that was not used on any of the original production reels or on any of the other edited reels.*

Once the final cut is approved, you will find yourself with an edit decision list that is comprised, for the most part, of time-code numbers from the recut master(s). Of course, to edit online with the original production tapes you will need to get back to the time code on those original tapes. In a complicated project with multiple dissolves, audio mixes and other effects, finding those original time code numbers could take days.

That's where the trace program comes in. Using it, an editor can find time-code numbers referenced to the original master tapes almost instantly. The editor begins by loading the edit decision lists from each cut (generation) into the editing computer. Each list should be assigned the same number as the workprint reel to which it corresponds, and, preferably, the lists should be cleaned through the 409 program before they are entered into the computer. Although the computer will ask for the reel number of each list, the lists do not need to be loaded in any particular order or sequence, except that the final version EDL must be loaded as reel ''F'' (final).

Once the lists have been loaded into the computer, the editor simply presses the ''L'' (for ''list'') key on the computer's keyboard. In a matter of seconds, the computer will transmit the final, fresh EDL to the line printer. A check of this new final (or ''F'') list readout will show that all play-in/play-out times refer back to the original master tapes and that all record start times are calculated to correspond to the final version list (see Figure 5.14). By pressing another button, the editor can load the new list onto punch tape or a floppy disk—whichever is appropriate for the editing situation at hand.

In 1980, in recognition of the impact the 409 and trace programs have had on the TV industry, Dave Bargen received an Emmy award from the National Academy of Television Arts and Sciences.

Edit Listing on the Personal Computer

The recent availability of low-cost personal home or office computers has made edit listing and edit list management a reality for video editors with limited equipment budgets. According to Lon McQuillin, president of McQ Productions and the originator of Edit Lister™, the Edit Lister™ program is available through Comprehensive Video Supply Corporation for Apple, IBM PC and IBM clones, as well as an MS-DOS version and CP/M for Kaypro and Sony computers.

Edit Lister™ provides video editors with a tremendous advantage by allowing them to prepare, store, revise, clean and transfer edit decision lists on their home computer instead of tediously hand logging or paying for computer time at an online post-production facility. In addition, when the personal computer is connected to an offline edit controller via its RS-232 edit output port, up to 995

```
REEL SUMMARY, SINGLE-SOURCE EVENTS

   REEL      COUNT

    20        001
   901        003
    21        000
   200        000

REEL SUMMARY, TWO-SOURCE EVENTS

    REEL1    REEL2    COUNT
```

Figure 5:13. The reel summary list from the 409 list cleaning program.

edits (depending on the computer model) can be automatically entered into the computer memory. McQuillin's Edit Lister™ features:

• Edit list cleaning for either A or B mode assembly.

• Re-edit with record time ripple.

• Adding and deleting edits.

• Adding post-production notes into the list.

• Including program effects such as dissolves, wipes and keys.

• Moving a single edit or blocks of edits.

• Changing source reel assignments.

• Renumbering the edit assignments.

• Edit over-record cleaning.

• Deletion of superseded edits.

• Split edit entry.

• Moving insert edits that bridge audio-video dialogue edits to the proper position for A mode assembly.

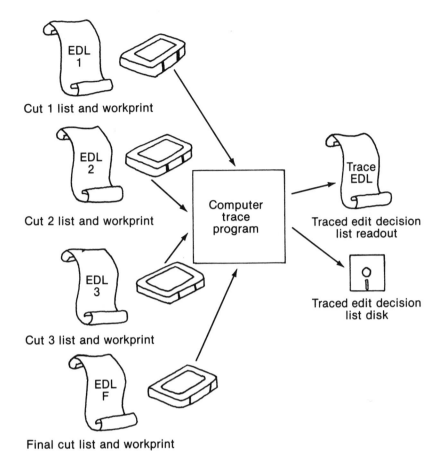

Figure 5:14: The trace program uses the edit decision lists from all offline workprints to generate a final EDL with time code numbers referenced back to the original production reels.

• Combining match frame edits into single edits.

• Identifying any gaps inadvertently included in the list.

• Outputting the edit list in either CMX, Grass Valley or Convergence formats.

An add-on feature is the Edit Lister™ D-LINK UP GRADE package which includes the necessary software, controller and shielded cable, and 8-inch disk drive. It enables edit decision lists to be stored on an 8-inch CMX compatible floppy disk, instead of the 3 1/2-inch disk or paper punch tape.

Editors using the Edit Lister™ program can output their lists onto 8-inch disk (via the D-LINK package), 3 1/2-inch disk, paper punch tape and even to lap type computers, depending on the online facilities capability. Future plans for the Edit Lister™ include a Macin-

tosh compatible program, a film feet and frames to television time code conversion edit lister program, and a trace program.

CONCLUSION

As I stated at the beginning of the chapter, the offline editing process can include a wide variety of activities, depending on the needs of the particular project and the preferences and predilections of the post-production team. Of course, the choice of an editing system, as well as the final design of an offline session, are matters decided by those in charge of the project. Regardless of the particular system or process used, the end products of all offline editing sessions are the same: a workprint that can be evaluated by producers and other key project officials and an edit decision list that can be used for assembling the edited master tape during the online editing process.

6 The Online Editing Process

Online editing is the stage of video editing in which shots, scenes and segments from the original source reels are pieced together to form the final edited master. This is analogous in film editing to cutting the negative, performing all opticals, completing basic sound editing and mixing, correcting color timing and creating the final release prints. The difference in videotape editing is that the editor performs *all* of these functions in one session. As a result, online editing is the most technically demanding phase of the video editing process.

The online session can also be the most stressful phase of video editing, for client and editor alike. Since online editing is the final stage of the editing process, it is the time when all earlier cost overruns come home to roost. With deadlines looming, it is also the stage when editors and clients begin to feel the full effects of earlier schedule overruns. Keeping this in mind, I always try to adhere to the first rule of videotape post-production: Be prepared. In other words, before you begin the online session, make sure you're ready.

MAKING SURE YOU'RE READY

Your production shoot is finished, and you've been through offline editing, so you're all set to go online. But are you *really* ready? Whether you're renting equipment or editing on your own system, it's best to make sure. In either case, improper preparation inevit-ably leads to higher online costs and an inferior final product.

To make sure you're ready, let's review a few of the factors discussed in earlier chapters. Before moving on to online editing, you should have in hand one of the following: detailed editing notes taken at pre-screening sessions or a completed and approved edit decision list (in hand-logged, punch tape or floppy disk form) from an offline editing session. Remember, too, that you should have already confirmed that the punch tape or floppy disk is compatible with the online editing system. For instance, an edit decision list stored on a 5 1/4-inch disk is worthless if the online system is equipped to handle only 8-inch disks.

You should also make sure that all of the source material is at the online facility *prior* to the scheduled start time. If possible, all film or audio transfers should be completed ahead of time to avoid costly delays and unscheduled equipment needs. Also, if the project includes source material recorded on several different formats (2-inch, 1-inch, 3/4-inch or 1/2-inch), the director and editor should confirm that the online system is equipped to accept the required formats. Along the same lines, the director and editor should make sure that the online system can handle mixed time code (drop-frame and non-drop-frame) interchangeably, if the project requires it. Finally, before editors and directors take the leap into online editing, it wouldn't hurt for them to take one more look at the preparations for post-production discussed in Chapter 4 and the offline editing process described in Chapter 5.

THE ONLINE EDITING BAY

As discussed in Chapter 3, online editing bays can range from the simple and utilitarian to the posh and elegant. Even inexpensive 3/4-inch editing systems can, and often do, function as online systems, as long as they are equipped to record and monitor a professional-quality edited video production. As Figure 6.1 shows, minimum equipment requirements for a professional-quality online editing bay include the following:

• An editing controller;

• Two playback VTRs with time base correctors;

• One record VTR capable of assemble and insert (audio/video, audio-only and video-only) edits;

• A video switcher with at least one special effects row;

• Waveform monitor and vectorscope display units;

• A video processing amplifier;

• A master color bar and sync generator;

• A black-and-white title camera;

• A small audio mixing board with equalization capability; and

• Audio and video monitors.

The online editing bay should also include sufficient room for both editor and client to move comfortably and provide adequate work space for scripts, notes and telephones.

Editing Bay Aesthetics

Fortunately, the days are gone when the professional-quality online editing bay was both machine room and control room—in other words, when noisy, large-format VTRs were located in the same room as the editor, director and other members of the editing

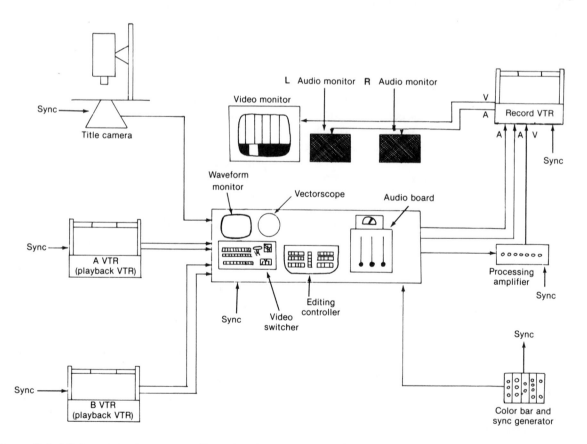

Figure 6.1: Minimum equipment configuration for an online editing bay.

team. Today, the editing team usually works in a separate, acoustically controlled editing "suite," from which all VTR functions are controlled with a single console. In fact, as editing bays have become more complex, system integration (the ability to control all equipment through the computer console) has become essential.

Editing suite aesthetics have also become increasingly important, particularly as clients and editors have begun to recognize the psychological impact that editing bay design can have during the long hours spent in online sessions. As I have already suggested, minimum design considerations include sufficient room size, adequate lighting (designed to hold down glare) and enough work space for the client's notes, scripts, etc.

Often, a relatively simple design decision can make the difference between a successful and a not-so-successful editing environment. For example, some editing bays are designed with such a total concern for equipment placement that the client's work space ends up being a small table located behind the editor's control console. As a result, the client must shout his instructions to the back of the editor's head—a situation that makes it difficult for the editing team to work in cohesive harmony. A better arrangement, by far, has the client's work space located next to the editor's control console, even if this requires relocating one of the less frequently used pieces of equipment.

I recall working in a particular edit bay that always seemed to make the members of the editing team feel tired and depressed. I finally realized that its ceiling was much lower than that of most other editing bays, resulting in a general feeling of claustrophobia. For readers who are in a position to design their own editing bay, I recommend paying attention to all of these environmental factors. For editors, the positioning of the equipment that surrounds the control console is especially important. The equipment should be arranged to minimize stretching and reaching—movements that can become very tiring when they are repeated hundreds of times during the course of an online session.

Of course, even with the most minute attention to detail, it is impossible to design a single editing bay that will meet the needs of every project. For example, for clients who are producing small-scale industrial training tapes, editing in an acoustically perfect post-production suite equipped with stereo audio mixing capabilities and multi-effects video switching equipment might qualify as technical (and budgetary) overkill.

Multiformat Videotape Editing

Until recently, source tapes from projects that were produced on different videotape formats needed to be transferred to a common format for time-code editing. In the process, a generation of video quality was lost.

CMX Corp. changed this when they introduced the Intelligent Interface Unit (I^2 for short) as part of their CMX 340 Computer Edit System. Using serial data signals sent through a three-conductor audio patch cord, the I^2 units convert the data to the individual parallel control signals for 2-inch quad VTRs, 1-inch B or C format VTRs, 3/4-inch videocassette VTRs, 1/2-inch videocassette VTRs, audio mixing boards, video production switchers, and multiple-track audio recorder/players.

With the advent of the Intelligent Interface Unit and subsequent serial direct control circuitry now available with some VTR makes, most major online editing facilities are now equipped to handle multiple videotape formats. As a result, it has become quite common for online sessions to incorporate source reels recorded on several formats—particularly as more production shoots are recorded on smaller format VTRs.

VIDEO PROGRAM FORMATTING

During the online session, both video and audio material are recorded onto the edited master tape according to a predetermined design or format. The exact format used in a given online session depends on a number of factors: the "house style" of the production company, whether the program is a corporate training tape or a network production, the manner in which the program will be transmitted, etc. Figure 6.2 illustrates a typical format for a network TV program. It includes the following components:

1. Approximately five seconds of blank tape at the head of the tape, for threading purposes.

2. At least one minute of standard video color bars and a 0 VU reference audio tone, both recorded under the exact conditions prevailing when editing of program material is performed.

3. An identification "slate" containing information such as the program title, production number, length of the program material and date of the edit session. A 10-second slate is sufficient.

4. Ten seconds of video black prior to the point where the program material fades up. In place of the

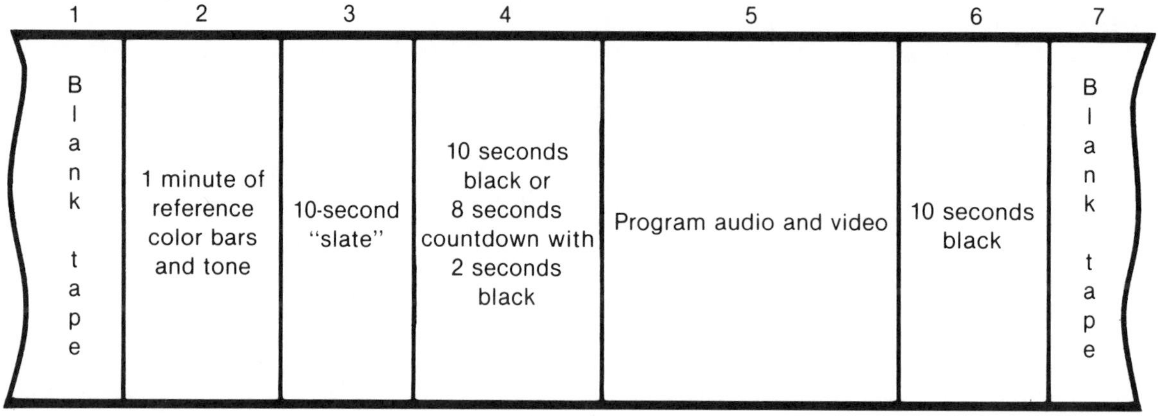

1	2	3	4	5	6	7
Blank tape	1 minute of reference color bars and tone	10-second "slate"	10 seconds black or 8 seconds countdown with 2 seconds black	Program audio and video	10 seconds black	Blank tape

Figure 6.2: The basic delivery format for network projects completed on videotape.

video black segment, some production companies include a visual countdown, starting at eight seconds and ending two seconds before the program video fades up.

5. The program material. This includes all of the edited program video, as well as the fade up from black and fade out to an ending logo. All commercial breaks, if there are any, are included in this segment. Figure 6. 3 shows a timing sheet used at Vidtronics, Inc. to lay out the program material portion of an hour or half-hour network program.

6. Ten seconds of video black, following the runout signal or final fadeout of the program material. This allows a smooth transition to other material when the program is broadcast or screened.

7. The program tail. After the runout signal, there should be 20 seconds of blank tape, left as a "tail," to prevent the tape from running off the end of the reel. I also try to leave one minute of extra blank stock on small 1-inch reels, to help prevent stretching from tape tension.

As a general rule, a 1/2-hour reel of videotape contains approximately 34 minutes of tape, a 1-hour reel contains about 64 minutes, and a 90-minute reel contains about 96 minutes. Consequently, there is usually plenty of tape left as a tail on the end of the reel.

SAMPLE ONLINE SESSION

No discussion of online editing would be complete without a description of a sample online session. Let's begin by describing our fictional editing bay. As online facilities go, it's fairly typical, consisting of a time-code editing system, a video switcher with one or more special effects banks, a video processing amplifier, an audio mixer board, at least one industry-standard color monitor and three or more videotape machines. This particular editing room is also part of a multi-bay post-production facility, which can offer less frequently used special equipment in certain editing bays.

Preliminary Check

Enter the editor. Since most editing bays are used for a variety of post-production projects, our editor should begin by checking the configuration of the equipment and editing controls. An editor should *never* assume that the bay was left in a normal configuration at the end of the previous editing session.

First, the editor should check the setup of the video processing amplifier on a vectorscope and waveform monitor, using system-standard color bars from the editing bay's video switcher as the video input. Is the setup correct (7.5% in NTSC or 0% in PAL)? Is video peak white (100 IRE units in NTSC or 700 mv in PAL)? Is the horizontal sync level correct (40 IRE units in NTSC or 33 mv in PAL), and are all color vectors properly aligned in the vectorscope display? In addition, each parameter of the processing amplifier that has a front-panel adjustment should be checked.

Next, the editor should check the control room's color monitors for proper adjustment, using color bars from the video switcher as the video input. Due to technical limitations inherent in color picture tubes, it is extremely difficult to get two color monitors to

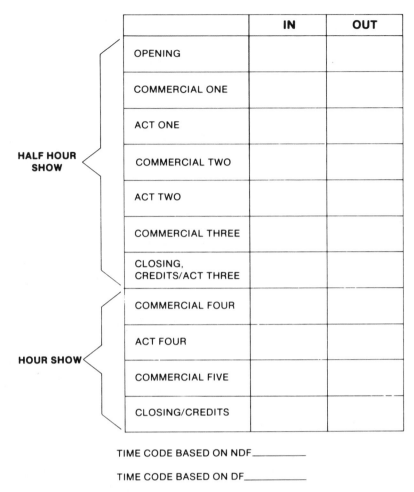

	IN	OUT
OPENING		
COMMERCIAL ONE		
ACT ONE		
COMMERCIAL TWO		
ACT TWO		
COMMERCIAL THREE		
CLOSING, CREDITS/ACT THREE		
COMMERCIAL FOUR		
ACT FOUR		
COMMERCIAL FIVE		
CLOSING/CREDITS		

HALF HOUR SHOW

HOUR SHOW

TIME CODE BASED ON NDF_____

TIME CODE BASED ON DF_____

If no time code is present please signify the tape counter time by preceeding the numbers with a TTT

*(Tape Counter Time should be calculated by resetting the counter to 0:00:00:00 at 1st frame of program video).

Figure 6.3: Timing sheet used for laying out the program portion of a network production. Courtesy Vidtronics, Inc.

look identical. In fact, many online edit bays contain only one program color monitor, to avoid the time lost trying to determine which monitor is "correct."

Additional Facilities

Perhaps our sample editing session requires additional equipment (1/4-inch audio machines, color cameras, electronic graphics equipment, other types of VTRs, etc.) that are not normally found in this particular editing bay. The editor should make sure that this additional equipment is available and

"patched into" the editing system. Typical patches include video and/or audio feeds, plus cords for transmitting control commands to the added equipment.

Initial Facilities Setup

Any additional equipment, plus the equipment normally housed in the editing bay, must be properly set up (or "phased out") before editing can begin. Phasing out is the process of adjusting the color bar signal recorded at the start of the original production reels (the reels that will be used as playback tapes in the online session) to conform with the color bars generated by the editing system's video switcher.

To begin the phasing out process, the editor should put the video processing amplifier (proc amp) in the "direct" mode. This means that the color bar signal from the video switcher will pass through the proc amp unprocessed, so the editor can make a direct comparison of the burst and horizontal sync components from the video switcher and playback VTR video signals.

By now, the editor's assistant should have loaded the first reels needed for the online session into the playback VTRs. Playing the reels from each of the playback machines in succession, the editor should perform the following adjustments for each VTR. First, using the effects row on the video switcher, he should prepare a dissolve from the reference black signal generated by the video switcher to the signal supplied by the playback VTR. As the editor moves the fader handle on the switcher from one source to the other, he will see that the display on the vectorscope switches from reference-black burst to the playback burst. These burst levels and phases should be identical. If not, the editor should instruct his assistant to adjust the playback VTR's overall phase control (also called the subcarrier control) and its burst level control, to bring the playback VTR into proper adjustment. Since time base correctors for helical VTRs do not have front panel controls for burst levels, extreme level adjustments may require internal alignment changes—an operation that is best handled by a maintenance engineer.

Any additional equipment (digital video effects equipment, color cameras, etc.) should be checked the same way. For example, color bars from the editing bay's video switcher should be fed through the digital video effects channels, so the digital effects can be phased out, following the same procedure used for playback VTRs.

Horizontal Sync Timing

Next, our editor should check the horizontal sync timing of the playback VTRs and any external equipment connected to the editing system. Working from the effects row of the video switcher, the editor should prepare a dissolve from color bars generated by the video switcher to the signal supplied by the playback VTR. If the horizontal timing is correct, the picture on the color monitor will remain stable and not jump horizontally as the dissolve is performed. If the picture does jump, the editor's assistant should adjust the

horizontal timing control on the playback VTR until there is no visible horizontal instability. Usually, this procedure works best when the editor moves the fader handle continuously while the assistant adjusts the playback VTR.

Once the horizontal sync timing is adjusted, the editor should return the video processing amplifier to the "normal" mode.

Record Machine Audio/Video Levels

The final step in the initial setup procedure is to check that the proper audio and video levels are being fed from the playback VTRs to the edited master (record) VTR. This involves feeding audio and video test signals into the edited master machine and checking the relevant meters on the VTR. Like all other initial setup procedures, this step needs to be performed only once, at the beginning of each editing session.

Playback VTR Setup

Playback VTRs must be set up each time a reel is changed. In this procedure, the editor adjusts the color test signal at the head of the new reel so it conforms with the standard test signal generated by the editing bay's video switcher. Although there are several ways to make this adjustment, I have found the following method to be the most consistently accurate.

First, the editor prepares a wipe on the effects row of the video switcher. The wipe should compare the color bars generated by the video switcher with the color bar test signal recorded at the beginning of the playback reel. Once the wipe is set up, the editor should adjust it so the control room monitor displays predominantly the video from the playback VTR. The video switcher (or "reference") color bars need only show enough for both images to appear on the vectorscope and waveform monitor screens (see Figure 6.4). Working from the vectorscope and waveform monitor displays, the editor can then adjust the signal from the playback VTR so it conforms with the reference color bars.

The audio test tone from the playback VTR should also be adjusted, to the point where the tone registers zero on the VU meter of the edited master VTR. Finally, if a project includes isolated camera reels, the shots from the reels should be compared against each other, so the editor can make any last-minute color corrections that may be necessary. With these final adjustments, our editor should be ready to begin editing.

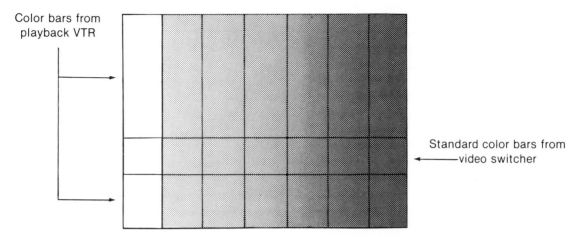

Color bars from
playback VTR

Standard color bars from
video switcher

Figure 6.4: Monitor screen set up to display color bars from the playback VTR and standard color bars generated by the video switcher.

ENTERING EDIT POINTS

As Chapter 3 pointed out, editors can choose from a number of techniques for entering edit points during the online session. For example, editors can opt for one of the manual entry techniques: marking in/out points "on the fly" as the tapes are played in real time or setting in/out points by typing time code numbers into the computer (and then changing those location points, using the "trim" functions of the editing controller). However, since the majority of online editing sessions are now preceded by some type of offline editing, the editor usually has an edit decision list (EDL), either hand-logged or stored on a data device, to be entered into the editing system for subsequent automatic assembly.

Many clients who log their decision lists by hand go on to reserve computer time (at a nominal fee) prior to the online session. They then manually enter the edit decisions into the computer, which gives them back a punch tape or floppy disk version of the list that can be used for automatic online assembly. This can be an effective and efficient method of preparing for an online session, as long as the client is reasonably confident that his hand-logged list is accurate. An inaccurate hand-logged list can transform the online session into a list management exercise—and a waste of the time that should have been saved by going to auto assembly.

Manual Entry of Hand-Logged Lists

If the online session is one in which the editor is using a hand-logged EDL for manual edit-point entry,

I recommend having a second party read the list to the editor. In this way, the editor can concentrate on entering numbers quickly and accurately. For the crew member reading the list, here are a few hints that will help things run smoothly.

• Don't read "leading zeros" to the editor. For example, an entry point of 00:03:45:13 should be read as "3, 45, 13." Since the leading zeros are disregarded by most computer edit systems, the editor shouldn't waste time entering them.

• Read the list in a consistent pattern. For example, if you start by reading the reel number first and the type of edit second, continue with that pattern to the end of the decision list. The order you follow really doesn't matter, as long as the pattern remains consistent.

• Let the editor know when the time code numbers given for scene transitions refer to the middle point of the transition, rather than the beginning or end. With this information, the editor can automatically adjust the edit point while entering the data, avoiding the need for additional, and time-consuming, trim operations later.

Punch Tape or Floppy Disk Entry

As discussed in Chapter 5, punch tape and floppy disk systems both offer significant advantages over hand-logged decisions lists for fast, accurate entry of edit points. Today, most editors prefer working with floppy disks, primarily because of their small size and rapid data retrieval capabilities. Loading edit decision lists from floppy disks is a straightforward process—as long as the disk is compatible with the computerized online system. Working from the editing system's keyboard, the editor simply asks the computer to transfer the edit decision list from the disk to its memory (usually by pushing the keys marked "load edits" and "enter").

On most systems, the editor can load several sequential decision lists at one time, as long as event numbers aren't repeated. (If the numbers are repeated, they can be changed by activating the renumbering feature on the edit system controller.)

AUTOMATIC ASSEMBLY

With the EDL loaded into the editing computer's memory, our editor would be ready to begin automatic assembly. As discussed in Chapter 3, there are two different modes of auto assembly: A mode (also called "linear" or "sequential") and B mode (also called "optimum" or "checkerboard").

A mode is the simplest method of automatic assembly. Once the decision list is in the computer's memory, the editor's assistant simply loads the proper reels onto the playback VTRs and gives the system an instruction to start assembling. The system will continue to assemble the project in sequential order until it encounters an edit that requires a new playback reel (a reel that is not currently loaded on one of the playback VTRs). At this point, the computer will halt the assembly and ask for the new reel by its assigned reel number.

B mode is the more efficient form of assembly—as long as it is set up properly. In B mode assembly, the system performs every edit on the edit decision list that can be completed using the reels that are currently loaded on the playback VTRs (with the exception of transition edits in which only one of the required reels is loaded). As a result, B mode assembly reduces the amount of online edit session time needed for changing playback reels. However, to use this mode of assembly, the editor must make sure that the following conditions prevail:

• The edit decision list must have no edit time overlaps;

• All time and length factors pertaining to the project should be final before assembly begins;

• The preblacked master stock should be recorded on the edit VTR just prior to assembly; and

• Color setup of production material should be properly matched and not readjusted during the assembly.

I recommend B mode assembly for any projects that comply with these conditions, with the possible exception of projects that require numerous video special effects or multiple-reel transitions. Generally those projects can be edited more efficiently using A mode.

Many computerized editing systems now feature a function called "Lookahead." Lookahead means that the computer scans ahead on the edit decision list and pre-cues reels to the proper location for upcoming edits. On longer editing projects, the time saved through this pre-cueing process can be significant.

SPECIAL CONDITIONS AND CONCERNS DURING ONLINE EDITING

Of course, in online editing, there are a number of conditions and concerns that may require special attention.

Match Frame Editing

One of the great strengths of computerized editing systems is their ability to perform match frame edits. Also known as "frame cut" or "in frame" editing, the match frame process allows an editor to perform dissolves and other transition edits quickly and accurately, with no visible interruption or instability in the video signal. Previously, the editor had two choices. He could roll several VTRs in unison, or he could build an A/B roll (and thus lose a generation of video quality).

As shown in Figure 6.5, the old A/B method required recording the master audio track on both the A and the B reel for timing purposes, then editing the alternating scenes (with frame overlap) onto each reel, in sync. A member of the post-production crew would then roll both reels (using the master audio track as a sync reference) while the editor performed the transition manually through the video switcher.

Today, using the match frame process, the editor simply enters into the editing computer the type of transition, the playback VTR that contains the incoming scene, the corresponding time code locations and the duration of the transition. The computer takes over from there, automatically synchronizing the VTRs and performing the transition.

With computerized editing systems, editors can also

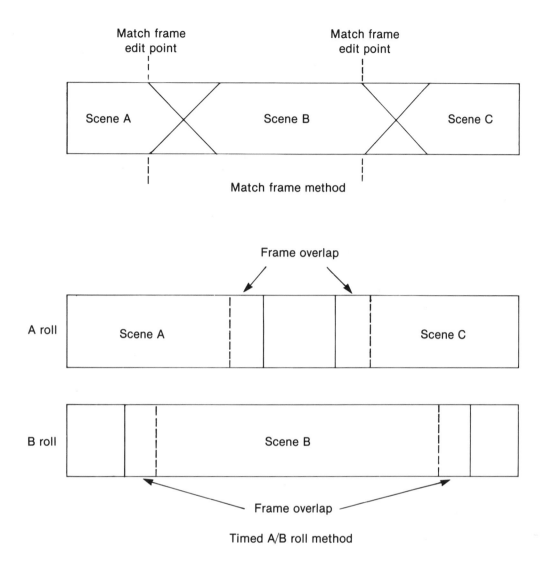

Figure 6.5: Match frame versus timed A/B roll editing.

perform long title key sequences without resorting to title rolls. Each title can be keyed in and out individually. The computer performs the edit and automatically updates the time code to keep the background scene in sync.

On computerized edit decision lists, match frame edits appear as the source and record-out times of the event preceding the transition event. For instance, in Format A of Figure 6.6, the source-out time listed as 01:20:32:00 provides the computer with the match frame reference for starting a dissolve of 45 frames with the incoming scene at 01:25:55:00 on the ISO-B reel.

Format B shows an alternative listing for the same dissolve transition. The first line of event #2 is a

zero-duration edit of the type used on editing systems that require "tracking edits," such as the Interactive Systems Co. (ISC) Super Edit system, CMX editing systems and others. These systems use what is known as a "passive list." That is, each event is stored in memory until it is needed, when it is transferred into the active portion of the system. In contrast, systems that use an "active" edit decision list keep the entire list in the active portion of the system at all times.

Editing with Mixed Time Code Formats

Many of today's more sophisticated computer editing systems can accommodate "mixed" time-code editing (using both drop-frame and non-drop-frame

| 001 | ISO-A | B | | 01:20:10:00 | 01:20:32:00 | 01:00:00:00 | 01:00:22:00 |
| 002 | ISO-B | B | D45 | 01:25:55:00 | 01:26:30:00 | 01:00:22:00 | 01:00:57:00 |

Format A

001	ISO-A	B	C		01:20:10:00	01:20:32:00	01:00:00:00	01:00:22:00
002	ISO-A	B	C		01:20:32:00	01:20:32:00	01:00:22:00	01:00:22:00
002	ISO-B	B	D	045	01:25:55:00	01:26:30:00	01:00:22:00	01:00:57:00

Format B

Figure 6.6: Different methods for listing match frame edits.

coded material simultaneously). However, since these systems require presetting (or "initializing") for one type of time code or the other, editors must be aware of the possibility of duration time errors. For example, suppose you are editing original material recorded in non-drop-frame time code onto a drop-frame edited master reel, with the editing system initialized in the drop-frame mode.

If, after performing this edit, you try a transition edit to another shot, you will probably end up with a "jump cut" on the match frame edit—the result of the time difference between drop-frame and non-drop frame code for the duration of the previous edit. In most cases, the timing error will be at least plus or minus two frames. To correct the problem, the editor will need to trim the match frame time code number on the source material by the appropriate number of frames. I don't recommend offline editing with mixed code formats because the problem of duration time errors would create chaos during online assembly.

Master/Slave Editing

The master/slave feature on online editing systems allows editors to control more VTRs than the three machines that are normally used in performing transition edits. By properly assigning the VTRs, the editor may roll, simultaneously and in sync, as many VTRs as the system can handle. The master/slave feature is found on most of today's more sophisticated editing systems, including Grass Valley Group's Super Edit system.

To use the master/slave function on the Super Edit system, the editor begins by activating the master/slave key found on the computer console. Once this key is pushed, a question displayed on the computer screen will ask the editor to assign a machine as a master VTR. As soon as he does this (by pressing the appropriate VTR select key), the system will ask the editor to designate a VTR as the slave machine. This time, after the appropriate VTR select key has been pressed, the system will ask the editor if he wants to assign another VTR. If additional slave VTRs are needed, the editor repeats the process designating the previous VTR selected as the master and assigning each master its own slave VTR. This is known as "daisy chaining" the VTRs.

The editor can also designate multiple record VTRs. On most systems, this is done by assigning another VTR as the record VTR's slave and then entering a numerical code to indicate whether the slave VTR should actually function as a record machine or simply play back in sync with the record VTR.

Sync Roll Editing

Sync roll editing (also called "sync mode" or "real-time" editing) is an offshoot of the master/slave process. In sync roll editing, all assigned VTRs are rolled in sync, and the editor switches "live" between shots available on the various reels. This method offers editors the ability to cut a scene as if it were live (using the computer console keys) and to receive a frame-accurate edit decision list in the process.

Sync roll editing works well on multicamera productions that were recorded on isolated cameras and on productions that were recorded in one continual sequence (concerts, plays, "live" situation comedies, etc.). Although dissolves and wipes cannot be performed during the sync roll process, they can easily be added later. Generally, this is done by using the list-management feature of the edit system to convert the simple, straight edits performed during sync roll editing to the desired type of dissolve or wipe transition.

Editing into Previously Edited Material

Many times it is necessary, for a variety of reasons, to edit into an existing edited master reel. For example, on commercials or syndicated programs, it is sometimes necessary to make title changes, to prepare different length masters and to insert audio tags. Often, editors must decide whether these edits should be performed on the master reel or whether it is better to perform the edits on a copy of the master (even though this would result in losing one generation of video quality).

On productions that will be run in a number of different markets or viewing situations (each of which might require changes in titles, program length or audio tags), I usually create a "generic" master—a master that contains none of the material that will be subject to change. Then I add changes by editing down a generation, creating what is actually a series of "custom-tailored" masters, one generation removed from the original program material.

When it is necessary to edit into the actual edited master, the post-production crew must make sure that the audio quality and video signal characteristics of the new material match those of the master material. The crew must also consider record machine optimization, tracking compatibility and color framing.

Matching the audio quality of audio inserts requires the careful matching of both volume and equalization characteristics, especially when the inserts are being added to a master that has been through the audio sweetening process (explained later in this chapter). Generally, I "dupe off" the section of sweetened audio being replaced, leaving several seconds of overlap at both the in- and out-points. Then, instead of cutting into an equalized track, I perform a short audio dissolve (10 frames or so into and out of the newly inserted audio). This softens the audio inserts, so they are far less apparent when the master is played back.

Matching the video characteristics on previously edited material requires careful attention to detail. Test edits should be performed comparing the standard color bar signal from the editing bay's video switcher to the color bars found at the head of the edited master tape. The editor must be sure to adjust the video level of the input signal so it matches the level of the test signal already present on the master videotape. Finally, before actually cutting into the master tape, the editor should make a test recording and compare it with the original material. Using the video and chroma level controls on the processing amplifier, along with the input level control on the record VTR, the editor can adjust the video signals so they match. Often, it is also necessary to "reoptimize" the record VTR.

Proper Color Frame Edits

In video editing, color framing refers to the phase relationship of the incoming signal's reference color burst to the reference color burst recorded on the edited master. The color framing is said to be "in phase" when the difference between the two is 0°, give or take a small amount. If the difference is 180°, the two reference color bursts are said to be "out of phase."

You can make this comparison by viewing a waveform monitor that is displaying the color reference burst of the edited master tape. Perform a test edit, and then keep your eye on the waveform monitor as it displays the first cycle of burst from the demodulated (unprocessed) video signal. As the edit plays through, the first cycle of burst should *not* change direction (see Figure 6.7).

If the first cycle of burst does reverse direction, the signal is out of phase. On a TV monitor, an out-of-phase edit will appear as a horizontal shift at the edit-in point and a whip (instability) at the edit-out point. You can correct 50% of the problem by running the playback signal through a processing amplifier, since a proc amp inserts new color reference and sync signals that are constant and extremely stable.

It is important to note that, whenever a VTR rolls and locks up, the color reference signal has a 50% chance of locking up in phase. Using the proc amp will maintain the phase of the incoming video signal as a constant. However, the record machine can still lock up out of phase with the incoming video signal, so editors must still be sure to monitor and correct problems when they appear.

Most computerized editing systems have their own color framing correction systems. With this feature, the editor doesn't have to keep checking color framing on every edit. Once a proper color frame edit is made, the computer will automatically correct the framing for the rest of the edits.

ADDING TITLES AND GRAPHICS DURING VIDEO EDITING

The process of adding titles to videotape is surrounded by a variety of myths and misconceptions.

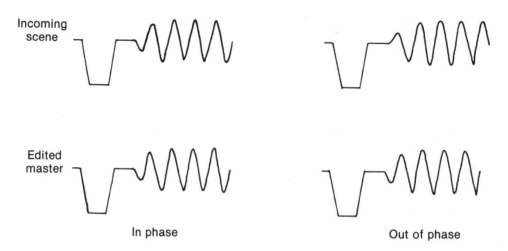

Figure 6.7: Comparing the phase of reference bursts to determine proper color frame edits.

For example, it's not uncommon for an advertising executive to show up at a post-production facility with a title card in one hand and a videotape copy of a commercial in the other. The executive is often surprised to discover, after a quick conversation with the post-production supervisor, that he can't simply ''add'' the titles over the scene. This misconception stems from the quite common belief that titles are merely ''burned into'' a previously recorded scene.

The Key Process

Our commercial advertiser would do well to memorize one of the fundamental precepts of videotape post-production: When a VTR is placed in the record mode and the videotape begins to roll, *all* video information that was previously recorded on the tape will be erased, to be replaced by the video signal that is currently being fed into the VTR. For example, our advertiser might try adding a title card to an existing videotape scene by feeding the signal from a graphics camera into a VTR, cueing the commercial to the appropriate scene and pushing the record button. However, upon playing back the tape, he will find that there now exists a nice picture of a title card but no background scene, since that was erased during the recording process.

What the advertiser actually wants to do is to ''key'' the titles over the existing scene. There are two ways to do this. He could make a duplicate copy of the master tape with the title added during duplication. Or he could take the original unedited production foot-

age (which, of course, he didn't bring with him) and reinsert the original scene, with the title added, into the edited master.

The second option is usually the preferred approach, since it does not involve losing a generation of video quality. It requires having the original production footage on hand and then ''keying'' the title into the scene during the reediting process. As described in Chapter 3, keying involves using a video switcher to combine the background graphics scene (cued up on the playback VTR) with a title card (set up in front of a graphics camera) and feeding the combined (''composite'') video signal to the record VTR (see Figure 6.8). Remember, though, that the keying is actually performed at the video switcher/special effects generator, with the record VTR simply standing by to accept the composite video signal (see Figure 6.9).

Post-Production Graphics Techniques

Adding videotape graphics can be a frustrating and time-consuming process. Once again, proper preparation can make the difference between success and failure. For example, all graphics and titles, whether they are generated electronically or prepared on title cards, should be approved, on hand and set up in their proper position before the final assembly session begins. Nothing wastes more editing time and money than searching for and making decisions about graphics during the online process. Of course, there are times when such decisions can only be made after viewing the edited scenes, as is often the case when

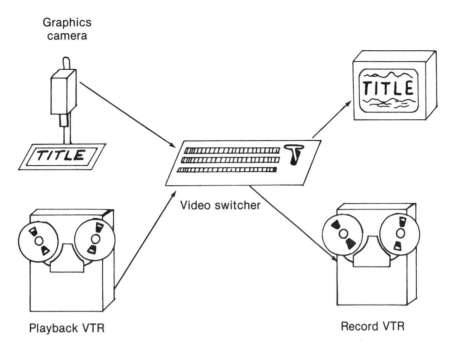

Figure 6.8: Keying process for videotape titles.

Figure 6.9: Titles inserted into a video image. Photo by Darrell R. Anderson.

the scenes involve digital special effects. As a general rule, however, titles and graphics should be prepared and approved well ahead of the final assembly session.

Of course, all graphics must fit within the readable confines of the television aspect ratio. In addition, TV graphics should not contain fine-line detail in or around the lettering. Finely detailed lines will normally not key properly, causing rough edges and "tears" in the video signal.

Normally, graphics drawn on art cards should be prepared using white letters on a black background. If the graphic is drawn as black on white, the editor should make sure that the title cameras at the post-production facility can be electronically reversed to accommodate the change. The video switcher will use the highest luminance level of the video signal for its key source, so the part of the graphic that is supposed to show up as a key has to contain the higher luminance level.

For a proper title key, the lighting should be uniform across the card. The angle of the light is not critical, but the lights should be arranged to avoid reflections into the camera lens. In most situations, the preferred method is to "wash" the art card, using lights on both sides of the camera. This diffuses the illumination,

creating a flat, even look across the graphic.

Copy stands such as the ones shown in Figure 6.10 can come in handy for shooting many types of art card titles. The camera can be raised or lowered and the art card carefully positioned to suit the particular shot. I recommend positioning the copy stand away from any overhead lights, since they tend to introduce unwanted reflections into the camera lens.

From my own professional experience, I know that the heat from lights can cause graphics to warp and curl, especially around the edges. For this reason, it often makes sense to pin graphics to a pegboard and to hold the edges down with a strip of aluminum, blackened to eliminate glare. In a pinch, the edges can be held down with any thin, heavy object.

Unless the camera is positioned exactly perpendicular to the graphic, a phenomenon called "keystoning" can become a problem. In keystoned images, one side of the graphic appears larger than the other and in slightly different focus (see Figure 6.11). This can be easily corrected by realigning the camera or the copy stand.

To create a good quality key, the editor must pay attention to the video level of the graphic. Figure 6.12 represents the waveform presentation of a properly

Figure 6.10: Copy stands for shooting title cards. Photo by Darrell R. Anderson.

Figure 6.11: Title keystone effect caused by the misalignment of the camera lens relative to the title art card.

adjusted keyed graphic mixed with the background video signal. Notice that the graphic video level is set at 100 IRE units (peak white). If the graphic video level is set lower than 100 IRE units, the keyed graphic will appear a dull gray against the background video on the TV screen. With most video switchers, graphics with video levels of 50 IRE or lower will not key at all.

In the process known as "chroma keying," color and luminance information are used as the key signal.

As discussed in Chapter 3, the chroma key process takes a uniformly colored background as the key source and then replaces part of that colored background with the video signal from another scene. Any primary or secondary color can be used for the key signal, as long as the lighting is extremely flat, producing a uniform background without texturing or shadows. However, the image that is being keyed into the new background should not contain any of the key

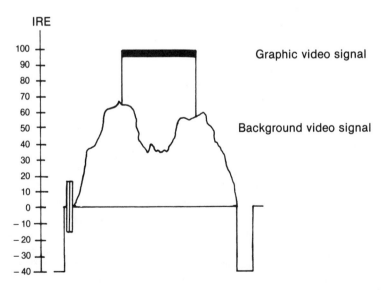

Figure 6.12: Waveform monitor presentation showing the proper key signal level (100% peak white) mixed with background video.

source background color, since that portion of the image will disappear along with the background. For example, a performer against a red key source background should not wear a red jacket. If he does, the jacket will mysteriously disappear when the key is performed. Blue is by far the most popular choice as a chroma key background source, primarily because skin pigmentation does not contain blue.

Electronic Graphics

Since their introduction in the late 1970s, electronic graphics systems have experienced tremendous growth. Born as simple character generators capable of creating letters in a single font, electronic graphics equipment has evolved into today's sophisticated systems that offer several font styles, multiple colors and animation features.

Compared to art cards, electronic graphics were at first thought to appear cheap and artificial. However, as the equipment has improved, electronic graphics have become widely accepted as the preferred option. In contrast to art cards, electronic graphics are fast and easy to change, and they provide perfect leveling and centering parameters.

Systems by Chyron Corp., Telemation, Inc., Aurora Systems, Dubner Computer Systems, Inc., and MCI/Quantel are examples of state-of-the-art graphics generators and animators. Most of these are capable of generating both single- and multiple-line titles, as well as title crawls, in white, colored and bordered font styles. Some systems (including the Aurora system) feature the ability to still-frame the video, to "paint" the frame electronically and to draw animated sequences electronically.

The primary disadvantage of electronic graphics systems is their cost. The most sophisticated systems can cost anywhere from $35,000 to $135,000. This compares to an approximate combined cost of $2000 for a graphics stand and black-and-white title camera.

VIDEOTAPE EDITING IN PAL

For the most part, NTSC and PAL editing procedures are very similar. The major difference is in the color framing process. Due to its phase alternation characteristics, the PAL video signal (see Chapter 2) has four different color frame phases, rather than the two phases found in NTSC. As a result, color frame phase between the edit master base recording and the original material may only coincide on every other frame. This, in turn, makes it impossible to edit PAL video frame-for-frame.

Recently, a colleague of mine, Harvey Berger, had the opportunity to edit a musical special in London on PAL editing equipment. To perform the musical edits required in the show while simultaneously cutting the video, he found that it was necessary to resort to split edits. The split system delayed the video edit one or two frames, effectively overcoming the color-frame error factor.

Some computerized PAL-format editing systems are programmed to flash a message on the screen before the edit whenever improper color frame phase is apparent. The computer then asks the editor if he wishes to correct the color frame phase. If the editor answers "yes," the system compensates for the phase error by advancing the playback VTR one frame.

Another possible problem in PAL editing involves edit list repeatability. If an edit decision list has been prepared through the offline editing process, it may not be possible to assemble the edited master from the time code numbers on the list, since the color phase of the edited master base recording may not (and frequently does not) coincide with that of the workprint base used in the offline editing session. To correct for this situation, it is necessary to adjust the "problem" edits by one frame.

WRAPPING UP THE EDITING SESSION

In online editing, as in many endeavors, "the job's not over until the paperwork is done." First, the editor must make sure that all of the reels used in the online session are properly labeled. Projects have a habit of coming back for changes or revisions, and it is an unlucky editor who must wade through unmarked reels months later trying to locate the footage needed for making adjustments.

Second, all edit decision lists should be saved and stored. I use a file cabinet to store my old lists for extended periods, since clients have called me as much as three years after a major project to request my lists and notes.

Purchase orders and work orders must also be filled out, to identify which equipment was used and for how long, and to record the total amount of videotape stock used (including both small "prebuild" reels and the final master stock). When completing these forms, an editor should try to be as clear and concise as possible. Responding to inquiries about a bill while referring to incomplete or inaccurate records can be a taxing and embarrassing experience for an editor— and a revenue losing experience for the post-production facility.

The final step in wrapping up an edit session is to restore all equipment controls and functions to their "normal" condition. All abnormal audio, video or control cables should be disconnected, and the level controls on the processing amplifier and VTR should be returned to their normal settings.

CONCLUSION

As in all aspects of video editing, careful preparation is essential to success in the online editing stage. Although a large part of that preparation involves adjusting and monitoring equipment, online editing is more than just a technical process. Producers and directors, as well as editors and engineers, will find that an understanding of concepts such as multiformat editing, automatic assembly, and techniques for adding titles and graphics is important for planning future projects.

Novice and aspiring editors must become proficient in dealing with these concepts. They must also be prepared to use their knowledge as a basis for dealing with the demands that inevitably appear with new generations of video editing equipment—such as digital video systems, which are the subject of Chapter 7.

7 Digital Video Effects

Fasten your seat belts. You are about to enter the world of high technology video, the world of digital video effects. Digital manipulation of the video signal has become the most used—and the most overused—method of creating and enhancing video special effects. To the TV viewer, these effects appear as the spins, multi-image montages and graphic zooms featured so insistently on almost every variety show, sporting event and news program.

In this chapter, I will describe how the methods for creating and inserting digital effects evolved. Then I'll look at the capabilities of five digital effects systems: The Vital Industries, Inc. Squeezoom and Magic, The Grass Valley Group, Inc. Kaleidoscope™, the Ampex Corp. Digital Optics System and the FOR.A VEC-440 digital effects controller.

ELECTRONIC DIGITAL FRAME STORES

A digital frame store consists of a network of microprocessor-controlled electronic chips, programmed to store an entire frame of video information—once that information has been converted from analog to digital form. This random access memory (RAM) network allows video professionals to alter the shape, size and position of the video signal by varying the reading (input) and writing (output) characteristics of the memory network.

Figure 7.1 is a simplified diagram of the frame store process. First, the analog video signal (the composite video signal discussed in Chapter 2) enters the analog-to-digital converter. In the converter, it is transformed into a digital signal—the only type of signal that can be read by (entered into) the random access memory of the frame store network. As the converted signal enters RAM, each individual frame of video information is held for a one-frame duration (1/30 of a second). During this interlude, the video frame can be altered and manipulated in a variety of ways. The altered signal then travels from the RAM network to a digital-to-analog converter, where it is transformed back into the composite video signal that can be recorded or displayed on standard video equipment.

It is worth noting that, with the one-frame delay, scenes with synchronized dialogue that are run through the frame store will end up with the dialogue out of sync. Fortunately, this is easily corrected during editing by delaying the corresponding sound by one frame.

Applications of Digital Frame Stores

Originally, digital frame stores were used as a means of correcting the large timing errors encountered in production situations where several remote (network feed, satellite feed, etc.) signals had to be synchronized and intercut during a broadcast. By converting the remote signals to digital form, the horizontal and vertical synchronization pulses of the signals could be replaced with pulses that were in precise sync.

A more recent, but little known, use of digital frame stores was to adjust the horizontal blanking signal on broadcast programs that were suddenly out of tolerance with the new Federal Communications Commission

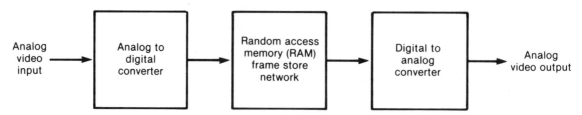

Figure 7.1: Simplified diagram of a digital frame store device.

guidelines (see Chapter 2). In fact, it's no secret that many post-production facilities paid for their fancy frame store devices by using them to correct their clients' out of tolerance programs.

Today, digital frame stores are most famous for their use in digital video effects systems. I first recall seeing digital effects on network sporting events, where scenes were zoomed and squeezed to form transitions between instant replay and live action sequences.

Commercial producers lost little time in putting the new video effects systems to a wide variety of uses. For example, before digital effects devices became available, those TV commercials produced on videotape that required the addition of offset video scenes (i.e., shots of the product) during post-production presented special problems. With no way of repositioning the offset scenes during editing, the scenes had to be shot in the precise position and framing needed to make them "fit" with the previously recorded portion of the commercial. I remember marking many color monitors with grease pencils and masking tape, in an effort to "line up" offset shots as they were being recorded. With digital effects systems, this alignment can be done electronically.

Of course, the ability to reposition offset scenes is only one of the many advantages offered by digital effects devices. With the capacity to zoom in or out, optically flip or tumble and roll shots, digital effects systems offer producers a great deal of flexibility to enhance video images during post-production. For example, using multiple-channel digital effects systems, producers can manipulate up to six channels at one time (that is, independently manipulate the size and shape of six separate shots on the screen) to create some dazzling visual collages.

Use and Misuse

Digital video effects are currently used in almost all facets of television production from industrial training tapes to network variety shows. Unfortunately, like any technical innovation, digital effects devices are subject to misuse. In all productions, digital effects should be used to enhance the on-screen performance, not to upstage it. For instance, digital devices are often used to frame a piece of news film beside an anchor during a broadcast. If the film "box" is carefully framed, this technique will generally enhance the newscast. But if the box is poorly framed or if it continually changes position, it will be more of a distraction than an enhancement.

On musical variety shows, directors often use visual effects around the singer to emphasize certain parts of the song. However, if the director is not careful, the effects can easily outdo the performer. On the other side of the coin, editors often encounter footage where the performance is so poor it needs to be enhanced or where scenes were so poorly directed they will not cut together properly. In such cases, a new version of an old film editor's adage is often appropriate: "If you can't fix it, 'effects' it."

FEATURES OF DIGITAL EFFECTS DEVICES

When they first encounter digital effects devices at a post-production facility, many clients ask the same question: What can those things do, anyway? The stock reply of many editors is, "They can do almost anything—you're only limited by your imagination." Actually, it would be more accurate to say that clients are limited by both imagination and budgets, since most digital effects systems are billed at a fat hourly fee.

On the pages that follow, I present an overview of the digital effects functions currently available by describing the features of five representative digital effects systems. Actual operation procedures are not given, as these are contained in each system's operating manual. Some diagrams of video patterns are also included as figures. However, readers should not assume that these provide a complete sample of the effects that are available on the systems.

Squeezoom

The Squeezoom, a trademark of Florida's Vital Industries, Inc., was the first commercially available multiple-channel digital effects system. Originally demonstrated at the 1977 convention of the National Association of Broadcasters (NAB), the Squeezoom was an immediate hit. It is currently available in one-, two-, three- or four-channel models; each channel on the multichannel systems can be independently programmed.

Figure 7.2 is a diagram of the control panel on Squeezoom's four-channel model. As the figure shows, this model features individual video input selectors for each channel, as well as controls for adjusting image size, aspect ratio, picture positioning, mirror imaging, upside-down imaging and freeze frames. Figure 7.3 illustrates the various video configurations that are available by manipulating the different effects buttons.

Operational features of the Squeezoom include the following:

• *Video compression/expansion*. Squeezes the picture from full frame to zero, or the reverse, either horizontally or vertically.

• *Freeze frame*. Freezes a full frame of picture information.

• *Zoom*. Reduces or enlarges the picture size from zero to eight times normal.

• *Video positioning*. Moves the picture horizontally, vertically or diagonally.

• *Aspect ratio*. Changes the picture aspect ratio from 4 x 3 to any other rectangular ratio.

• *Programmable patterns*. Allows the manual programming of effects as well as more than 100 pre-programmed effects.

• *Waveform and auto key tracking*. Senses the position and size of a chroma key signal or wipe pattern and fits the picture into the prescribed area.

• *Mirror imaging*. Electronically flips the picture along its vertical axis.

• *Upside imaging*. Electronically flips the picture along its horizontal axis.

• *Picture flips and tumbles*. Allows the picture to flip (on the vertical axis) or tumble (on the horizontal axis) either continuously or for a programmable number of times.

• *Manual or automatic effects transition*. Allows the operator to activate effects transitions manually (using the fader handle) or automatically (using the programmable automatic transition feature).

• *PSAS (optional)*. The production switcher automation system (PSAS) stores the Squeezoom effects functions for instant recall at a later time.

• *Matte selection*. Gives the user the choice of ten hues for generated color backgrounds.

• *Video synchronization*. Allows the system to accept up to four nonsynchronous video signals (remote signals, electronic news gathering video, satellite signals, etc.) without having to "gen lock" the main house sync generator.

• *Stand-alone operation*. Gives editors the option of operating the Squeezoom independently or in conjunction with a video switcher and computerized editing system.

The Squeezoom's output panel features two composite video output jacks, for transmitting the combined output of all channels, and a separate external key output, for transmitting signals to the mixed effects rows on video switchers.

Magic

The Magic digital video manipulator, introduced in 1986, is a new product of Vital Industries, Inc.

Magic (shown in Figure 7.4) is a single-channel device that uses a simplified color touch screen and three-axis track ball for selecting operating modes. Magic features a built-in hard disk drive for recalling preprogrammed effects, as well as a micro floppy disk drive for pattern storage and entering software updates. Some of the operational features of Magic include:

• *Size, position and aspect control*. Allows for variation of image size, image position and image aspect ratios.

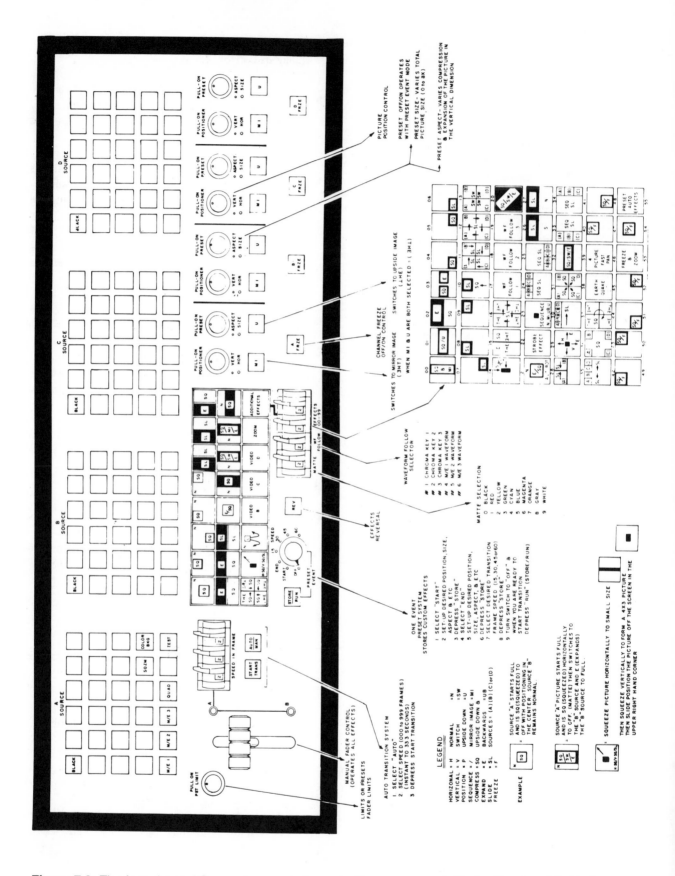

Figure 7.2: The four-channel Squeezoom control panel with programmed effects. Courtesy Vital Industries, Inc.

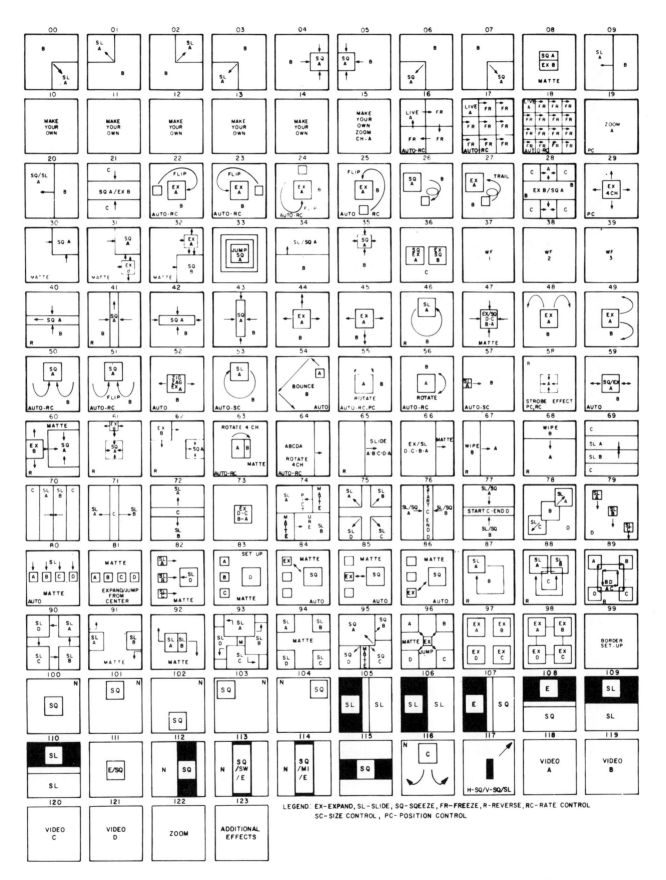

Figure 7.3: Video effects available on four-channel Squeezoom. Courtesy Vital Industries, Inc.

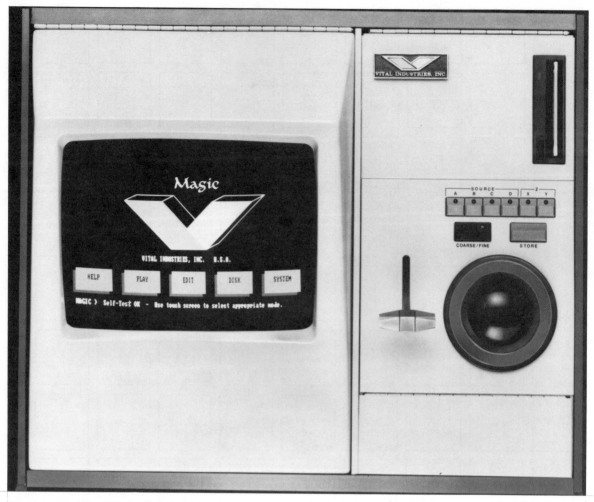

Figure 7.4:   The Magic digital video manipulator by Vital Industries, Inc. Photo courtesy Vital Industries, Inc.

• *Mirror image and picture inversion*. Electronically flips the image along its vertical or horizontal axis.

• *X, Y, or Z axis rotation*. The image may rotate about the center of the horizontal, vertical or Z-axis.

• *Variable true perspective*. Gives the impression of a 3-D image rotation in a 2-D picture.

• *Transition swoop*. Provides a non-linear transitional move from the effect start point to the stop point.

• *Picture cropping*. Allows the cropping of designated picture areas.

• *Star trail*. A transitional image repositioning that leaves a decaying trail.

• *Variable matte background generator*. Provides a full screen color background.

• *Manipulatable freeze*. Allows a frozen frame to be moved or otherwise adjusted.

• *Video blink*. Freezes the video in a repetitive manner to create a strobing or old-time movie effect.

Magic can operate on either NTSC or PAL video and can be triggered as a stand-alone device by all editing systems or interfaced directly with Vital's PSAS 3000 Production Switcher Automation System. It provides a key-out signal as well as a key-input signal for altered shape effects.

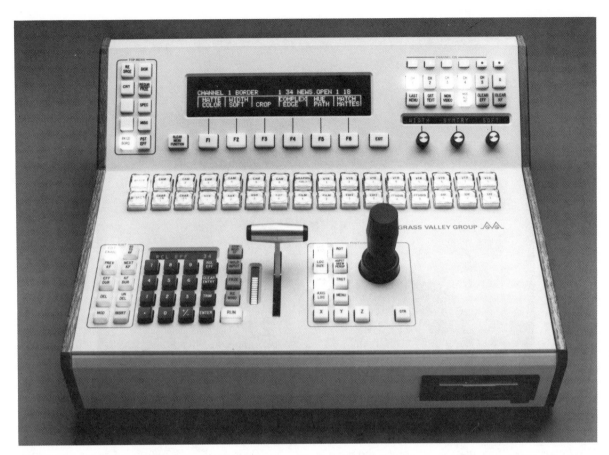

Figure 7.5: Grass Valley Group's Kaleidoscope™ digital effects system desktop controller. Courtesy of Grass Valley Group, Inc.

Kaleidoscope™

The Kaleidoscope™ DPM-1 digital effects system, manufactured by Grass Valley Group, Inc., was initially introduced at the 1985 SMPTE Convention, although marketing of the unit didn't begin until October 1986. The basic single-channel system consists of a control panel, controller and video processor unit.

The control panel is available in three versions. The first version is designed to integrate into the three mix/effects row model of the Grass Valley 300 series video production switcher (which we discussed in Chapter 3). The second version is designed to integrate into the two mix/effects row model of the Grass Valley 300 series video production switcher, while the third version is the stand-alone panel shown in Figure 7.5.

The standard control panel is designed for use with other types of video production switchers and includes the 3¹/₂-inch disk drive for preprogrammed effects, as well as the video source selector push buttons that are normally part of the 300 series video production switcher. The system is capable of storing 1000 key frames or a maximum of 100 effects per disk.

Kaleidoscope™ is designed to operate with either PAL or NTSC totally composite video, or a mixture of composite, analog component and digital component video. It can be configured (with the combiner option) into a system using a maximum of six channels and four separate control panels. Some features available on the Kaleidoscope™ include:

• *Translation*. Changes various analog input formats to a common digital format for interchangeability.

• *Rotation*. Permits picture rotation around all three axes.

• *Scale*. Creates proper size scale when effect is being rotated in the perspective mode.

Figure 7.6:   Video image representing the Kaleidoscope™ mosaic effect. Photo by Darrell R. Anderson.

• *Perspective*. Allows the operator to add apparent depth to the picture.

• *Perspective dim and fade*. Adds dimming, shadowing and apparent focus changing as effect recedes in distance; also adds light glare effect.

• *Multi-mode freeze*. Several selectable freeze points for strobe, decay or sparkle effect.

• *Contrast enhance*. Adjusts luminance gain without affecting chroma, image enhance, and fade out in distance.

• *Video reverse*. Gives color negative effect to normal video and normal video to color negative transfer from telecine when transferring effect directly from film.

• *Monochrome*. Provides black and white output.

• *Video borders and backgrounds*. Provides picture borders with adjustable width and color, as well as full background colors.

• *Mosaic*. Transforms the picture into size selectable mosaic squares (Figure 7.6).

• *Picture cropping*. Allows cropping of designated picture areas.

• *"Drop shadow" effects*. Creates an offset shadow matching the size and shape of the effect.

• *Input recursive memory*. Provides for noise reduction and motion decay.

• *Optional output recursive memory*. Enables creation of strobe titles (Figure 7.7), montage, multifreeze and motion decay.

• *Programmable effects*. Allows storing of preprogrammed effects on a 3 1/2-inch floppy disk for subsequent recall.

The output of the basic Kaleidoscope™ DPM-1 consists of one PAL or NTSC composite video and one set of component video (RGB or Y, R-Y, B-Y), plus the key signal output for use in video production switchers during external key situations.

Ampex Digital Optics System

The preliminary version of the Ampex Digital Optics (ADO) System was introduced at the 1981 NAB

Figure 7.7: Using the multi-mode freeze feature on the Kaleidoscope™ altered title effects such as that shown above can be achieved. Photo by Darrell R. Anderson.

Convention. It was the first digital effects system to offer image perspective—the illusion of depth in the two-dimensional picture. Since 1981, several advancements have been integrated into the ADO system resulting in three separate models: the basic ADO 1000, the middle range ADO 2000 and the high-end ADO 3000.

The ADO system relies on software to achieve its flexibility. ADO's basic control package (see Figure 7.8) consists of a control panel with a joystick lever and push button keys and a monitor used for viewing the system's data display menus. In addition, the monitor unit includes a mini-floppy disk drive for storing and retrieving programmed video effect sequences.

The system operator manipulates the video image by activating the proper control panel key and positioning the joystick lever to achieve the desired change. These changes are stored in the system memory for subsequent recall.

Each entered video change on the ADO system is called a keyframe. Since any effect requires both a start and stop point, an effects sequence actually re-

quires the programming of at least two keyframes.

The ADO system can handle up to 24 keyframes per effect sequence (24 changes of video size, shape or position in a continuous sequence). In addition, 36 separate effects sequences can be individually stored on each mini-floppy disk.

The ADO's ability to achieve true image perspective and to give apparent depth to a moving effect (see Figure 7.9) was its key to success. While operators previously could control image motion along both the horizontal and vertical axes, ADO permitted moving images along the so-called Z-axis—the axis that adds a feeling of apparent depth to the TV picture.

Features of the basic ADO system include:

• *Thirty preprogrammed effects with single key recall.* Gives ''on the air'' operators immediate access to effects which can be modified at anytime.

• *Optional picture perspective.* Allows the operator to add apparent depth to the picture. The ADO 1000 comes either as a two-dimensional version or a three-dimensional perspective version.

Figure 7.8: Control package for the Ampex Digital Optics (ADO) System. Courtesy Ampex Corp.

• *Video compression/expansion.* Enlarges or shrinks the picture from zero to more than eight times normal size.

• *Freeze frame.* Allows the operator to freeze and manipulate individual frames or fields of picture information.

• *Video positioning.* Moves the picture along the vertical, horizontal or the Z-axis, or along a combination of all three axes.

• *Blur.* Creates a minor misfocus effect on moving images.

• *Programmable effects.* Allows operators to store programmed effects on a mini-floppy disk for subsequent recall.

• *Mirror imaging.* Electronically flips the picture on either of the three axes.

• *Luminance reversal.* Creates a color negative effect to normal video, or creates a white on black image from a black on white image.

• *Controllable aspect ratio.* Allows the video image to be changed from the normal 3/4 ratio to any desired ratio.

• *Video borders and background color.* Allows the operator to create borders of adjustable size, width and color, as well as create adjustable full screen color background.

• *Posterization/solarization.* Allows operator to adjust the luminance, color or hue of an image to create effects such as a ''cartoon-like'' appearance.

• *Soft edge keys.* Allows the operator to produce seven degrees of softness to the edge around key sources.

Figure 7.9: Sequence showing the axes of dimension on the ADO system. Courtesy Ampex Corp.

• *Globals*. Allows operators of multi-channel systems to manipulate all channels at the same time.

• *Auto cube*. Enables a prebuilt cube effect.

• *Dual video inputs*. Allows operator to have a different video image for both the A side and B side of each video channel.

• *Picture cropping*. Allows the cropping of designated picture areas.

• *Horizontal and vertical grid lines*. Provides calibrated horizontal and vertical lines that can be used to level and center the image.

• *Optional digi-matte*. Provides two external key signal inputs. Allows objects to be ''flown'' in against a background image.

• *Optional RGB video input/output*. Allows flexibility for connecting to other studio equipment.

• *ADO/AVC integration package*. Integration package for Ampex video production switchers that allows effects to be called up and controlled from the video production switcher.

A major upgrade to the ADO 3000 is Ampex's Concentrator which handles and combines up to four channels. The concentrator allows the operator the additional advantage of:

• *Smooth digital keying*. Multiple effects edges line up without the need for timing adjustments.

• *Auto priority*. Automatically computes the over/under relationships of images in a 3-D picture position.

• *Manual priority*. Permits overriding priorities or mixing with the auto priority feature.

• *Transparency*. Varies the image effect appearance from opaque to invisible.

• *Dimmers*. Controlled dimming of any channel to the concentrator matte background.

• *Light source*. Permits automatic control of each channel based on its 3-D relationship to a fixed light source.

• *Matte background generator*. One or two independent matte generators.

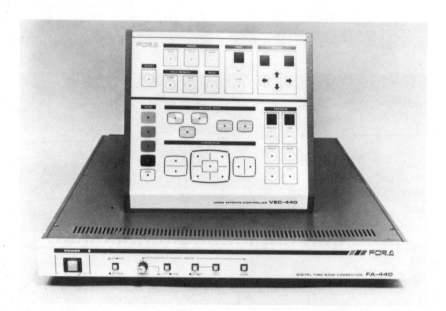

Figure 7.10:   FA-440 Digital Effects Time Base Corrector and Visual Effects Controller. Courtesy FOR.A Corporation of America.

Finally, the Infinity upgrade, using two or more ADO 3000 channels plus the concentrator, allows the operator to pass video frames, planes or objects right through each other. It also includes two frame store systems which give operators multifreeze and lets them create swirls, trails and sparklers in their effects—with or without video decay.

The ADO System is capable of operating up to four separate video channels with eight different control panels. Although the ADO can be set up as a stand-alone system, it is normally used in conjunction with a video production switcher.

FOR.A Video Effects Controller

The VEC-440 video effects controller, shown in Figure 7.10, is designed to be used as a companion with FOR.A Corporation's FA-440 Digital Time Base Corrector. It was created to provide editors with a wide range of digital video effects for under $10,000 (1987 price). This base price is several multiples below the price of the mainstream digital effects systems and gives educational, corporate and other nonbroadcast professionals a level of sophistication that was previously cost prohibitive.

This single-channel unit is compatible with most 1/2-inch and 3/4-inch VCR systems and provides high-grade time base corrector performance plus some of the most widely used visual effects.

Features of the Video Effects Controller include:

• *Video compression*. Permits half-screen or quarter-screen video compression displaying up to five frozen images at one time.

• *Posterization*. Permits seven step adjustment to produce "poster" or "cartoon" image quality.

• *Mosaic*. Permits seven step adjustment of horizontal and/or vertical mosaic squares.

• *Multiple motion*. Combines freeze-frame images with live action images.

• *Negative*. Reverses the polarity of the luminance portion of the video signal to create a color negative appearance.

• *Picture inversion*. Turns the live action video image upside down.

• *Color background generator*. Generates a selectable full color background when using compressed video images.

• *Memory*. Allows programming up to nine pages of 127 effects per page.

• *Automatic freeze frame*. Allows the picture to freeze automatically when the video signal is interrupted or severely degraded.

• *Strobe freeze*. Permits adjustable freeze duration to achieve a strobe or skip freeze effect.

The TBC/Digital Effects Controller package operates as a stand-alone unit or can be integrated with a video production switcher.

CONCLUSION

As the preceding pages suggest, digital effects devices are very versatile and complex video components. Thus, the sample listings of features I have offered here can only provide the most basic introduction to their design and capabilities. In fact, after some hands-on experience, many editors find themselves using digital effects systems to create innovative effects of their own—effects sequences that would surprise even the system's designers. However, I should pause to repeat the warning I gave at the beginning of the chapter: on all projects, care should be taken that digital displays do not simply become "effects for effects' sake."

It should be very apparent after reading the preceding chapters that the video post-production industry is in a constant state of change and improvement. We've seen tremendous changes in the design and capability of VTRs, offline editing systems, film transfer equipment, online editing systems and digital effects devices. Next we shall describe the practices and procedures in the post-audio area which used to be commonly referred to in video circles as audio sweetening. With the flurry of recent advancements in control sophistication and digital recording, it is now becoming referred to as audio post-production.

8 Audio Post-Production for Video

Audio post-production for video takes place when all the previously recorded audio sources are combined with additional sources, shaped, textured and mixed together to form the completed audio track. The term "audio post-production for video" is becoming the more appropriate term for the process instead of "audio sweetening," which many audio people think is not only outdated, but insulting!

The original term "audio sweetening" came about in the early days of video post-production when the one usable edited audio track was enhanced by adding some music, sound effects or narration. Adding in effects or music, not to mention any type of in-sync audio work, was a time-consuming and laborious process because of the lack of sophisticated control which today's state-of-the-art equipment makes possible (this predated the invention of SMPTE/EBU time code). Resolvers were available that kept the multitrack audio and VTR running at the same speed, but there was a problem with starting two or more devices simultaneously and locking up in sync. In fact, synchronization was the responsibility of the second audio engineer who would wear headphones to monitor the multitrack audio in one ear and the VTR audio in the other—any echo would indicate an out-of-sync condition.

Today's audio post-production equipment can accurately cue, synchronize, edit either frame or sub-frame accurately, digitize and synthesize completely new sound or music effects, and replace in-sync dialogue and sound effects. In short, audio post-production for video has come a long way from simply "sweetening"

in effects and music to enhance an already completed audio track.

Audio post-production for video allows the producer to create a wide range of aural impressions by manipulating the production audio using signal processing equipment, library or specially created sound effects, library or original music scoring and the talents of the audio post-production team. In fact, as we shall see later, there are network and cable film producers who prefer to complete their picture editing on video, but completely re-edit their dialogue audio tracks during audio post-production.

As a general rule, not all video projects require audio post-production. Many video projects which only require basic monaural dialogue mixing, audio equalization, some premixed music tracks and pre-recorded narration tracks, may just as easily be completed during the online edit session—providing that the online audio mixing console has the capacity and that the edit bay is acoustically adequate. In fact, some corporate and educational facilities are combining the editing and audio post-production process in the same room. However, it's really not cost-effective for a large broadcast post-production facility to equip its online editing bays to perform sophisticated audio mixing functions.

Experienced editors can usually determine if audio post-production makes sense for a project by assessing two factors: the complexity of the audio mix and the difference in cost of an audio facility versus the cost of the online facility required to do the mix during editing. Renting expensive audio facilities to add simple 1/4-inch narration tracks makes as much sense as paying for online editing time to mix a 24-track orchestration (assuming the edit bay's audio mixing console was even capable of handling 24 tracks). Both examples would constitute a tremendous waste of money and an inefficient use of the facilities.

Even though it might seem that the logical time for booking an audio post-production facility is after the online edit, this is not always the case. As we shall see later, most music video projects require a finished audio track for production playback and for video editing, whereas many commercials routinely produce a final mixed audio track and then use it to online edit the video.

UNDERSTANDING THE AUDIO POST-PRODUCTION MIXING ROOM

The same apprehension that afflicts many first time visitors to a video online edit bay also afflicts those experiencing audio post-production for the first time. At this point, I will reiterate my comment in Chapter 3 that "the whole is merely equal to the sum of its parts." By studying the various parts that make up the complete audio post-production mixing room we shall come to understand not only what the equipment is and does, but also the acoustical and aesthetic aspects that must be considered. (See Figure 8.1.)

Acoustics and Aesthetics

One of the first distinctive features of the audio post-production mixing room is the construction of the walls. The walls will be heavily padded or unevenly paneled with each wall constructed at a slightly different angle from the others. One or two walls may even be decorated with rough, porous rock which looks aesthetically pleasing, but which is mainly installed to disperse sound waves in different directions. However the interior walls are constructed, there should be no hard, parallel structures in the room that can reflect sound waves back and forth. Hard parallel surfaces set up a condition referred to as "standing waves." Certain audio frequencies (depending on the distance between the parallel structures) do not decay naturally with the rest of the audio spectrum, causing reverberations to be set up that change the acoustical

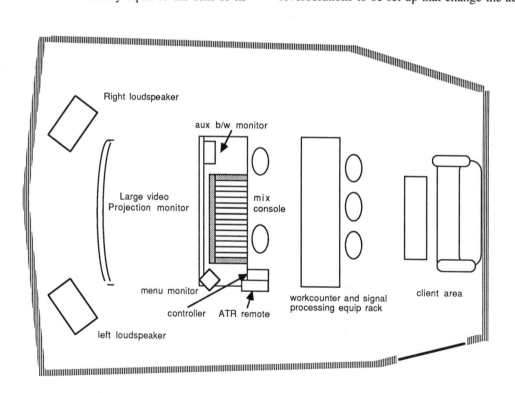

Figure 8.1: Diagram of a typical audio post-production main mixing room.

balance of the room. Simply, this means that what you are hearing is not what you are getting.

A second look around the room shows that as much equipment as possible is either recessed into padded alcoves or consoles, or is located in a separate room. This is done because metal equipment sides constitute hard, parallel surfaces and the equipment's fans and motors may generate some unwanted noise.

The acoustical engineering formulas that dictate the optimum dimensions of an audio mixing room are way outside the scope of this book. Practical considerations, however, such as the total number of people expected to occupy the room, the size of the mixing console, the amount of space needed for the signal processing equipment and physical building limitations should be included in the formula. In short, the audio mixing room should be large enough to work in comfortably and deep enough so that audio emanating from the speakers will not be reflected back to the intended listeners.

Aesthetically, the room can be as utilitarian or ornamental as tastes desire. Keep in mind that it won't help to design an acoustically correct room if it is decorated with hard surface furniture and pictures are hung on the walls. However, long, intensive hours can be spent in these rooms by the audio post-production team and clients alike, so it behooves anyone planning a mixing room to make it as aesthetically pleasing as possible.

The Audio Mixing Console

The audio mixing console is analogous to the online edit bay's video production switcher. Fundamentally, all of the source audio signals used during the mix are passed through the console where they are adjusted, manipulated and mixed with the other audio sources before they are sent back to assigned tracks on the multitrack recorder during the final mixdown.

At first, the mixing console's appearance is intimidating with its seemingly endless rows of switches, knobs and faders. However, when it is divided into its three basic sections—the input, the output and the metering/monitoring sections—the mixing console becomes more understandable. (See Figure 8.2 and Figure 8.3.)

The Input Section

The input section is identifiable because it contains identical rows of controls. In fact, the number of channels in the console is defined by the capacity of the input section because each complete input module equals one channel. Each module in the input section receives an audio signal from either a microphone, ATR, turntable, cartridge machine, cassette machine or any other audio source needed in the mix. It processes the signal, then routes it to the output section. Each input module may include:

• A *mic/line* switch for either a microphone or line input audio from ATRs, sound effect carts or other auxiliary equipment.

• A variable range *equalizer* for boosting or attenuating selectable frequencies of the input audio signal.

• *Prefader monitoring* controls for sending the input signal to either monitor speakers or audio effects devices.

• Two or more *program assign* switches to route each channel output to one or more of the console's submaster mixing buses.

• A *pan* pot (short for panoramic potentiometer) on stereo mixers for balancing between the left and right channel levels.

• A *peak level* indicator light and *trim* pot for adjusting the channel's headroom threshold level.

• A *solo* switch for sending the channel output to external headphones for monitoring.

• A *fader* control for providing accurate audio level adjustments either manually or with voltage controlled amplifier (VCA) automated fader controls.

The Output Section

The output section receives the processed signals from each input module (channel) and routes them through the submaster mixing buses to the submaster outputs and the monitor section. Most mixing consoles will vary in the amount of submaster outputs available. Some have either two, four or eight submaster outputs, in addition to the master monaural output which is a sum of all of the submaster outputs. The output section of the mixing console also contains the necessary switches, knobs and faders to control the submaster output levels, the master fader output levels, the calibration tone oscillator and the metering and monitoring bridges.

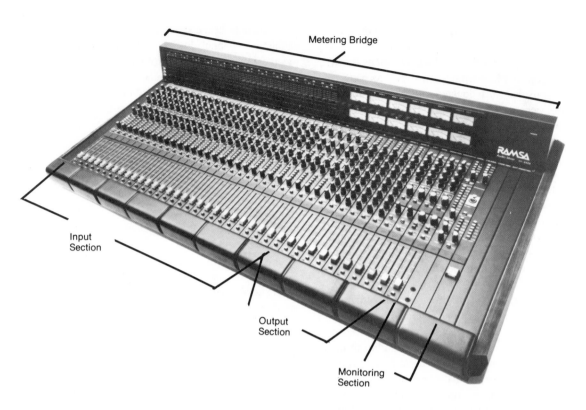

Metering Bridge

Input
Section

Output
Section

Monitoring
Section

Figure 8.2: The Ramsa Model-WR8428 28-Channel Audio Mixing Console. Courtesy Panasonic Industrial Co.

Figure 8.3: The Sony MXP-2012 Audio Mixing Console contains up to 12 manual fader channels and four VCA fader group outputs. Courtesy Sony Communications Products Company.

The Monitoring Section

The monitoring section receives the signals from the input section, the output submaster and the master buses and routes them to the metering bridge, the monitoring bridge and the solo output.

The metering bridge consists of either LED bar graph indicators, VU meters or a combination of both. These meters indicate the audio levels for each individual channel and submaster output, as well as the master monaural sum output and, depending on the console model, any other selectable outputs.

The monitor bridge is used to direct the selected monitor output to the main mixing room speaker system and any additional auxiliary speakers.

The solo output is useful for sending any selected input signal or combination of signals to external monitoring headphones which may be used in ADR, foley, music recording or any other use where individual or separate combinations of the overall signal must be heard separately.

The Editor/Controller

Having the most elaborate mixing capabilities available in audio post-production for video will not help your production if the equipment cannot be accurately controlled. There are a variety of controllers now available, each featuring its own version of the ultimate in audio editing. Some resemble video editing keyboards while others fit into the palm of your hand. Most of the controllers will control and synchronize the selectable master ATR, three or four slave ATRs and the playback VTR. All of the traditional tape transport controls are provided as well as programmable "multiple function" buttons which control sequences such as automated "looping" modes for ADR work or any other frequently used sequential functions. The controller also performs audio edits with up to one hundredth of a video frame accuracy (although the facility's ATR must be capable of performing clean edits with the same accuracy).

Systems like The Boss™, shown in Figure 8.4 also have the ability to store and recall audio edits from a floppy disk, list library effects, and display sound effect amplitude graphs on their color CRT menu monitor. CMX's new Computer-Aided Sound System (CASS-1) controller, shown in Figure 8.5, also becomes a mix automation system when it is interfaced with audio mixing consoles using VCA faders. It can memorize the motion of up to 32 faders and recreate the mix exactly as it was first performed, by entering start and stop time codes. CASS-1 can also store and recall several hundred mix versions with its 20mB hard disk. In short, audio editing controllers are becoming more sophisticated than video editing controllers.

External Signal Processing Equipment

In addition to mixing several audio tracks and accurately controlling the ATRs, we also need additional audio processing devices for either repairing problem audio sections or to enhance the audio track by adding or creating customized sounds. The complex structure of day-to-day sounds, as opposed to pure sine wave tones, gives the audio track its unique and vital quality. The tonal structure of each sound combines many varying pitches and volume levels in addition to individual blends of attack, sustain and decay. In audio post-production we can change these sounds into entirely different sounds by brightening, dulling, widening, narrowing, bending, delaying, strengthening, weakening, doubling, repeating, squeezing or expanding them by altering one or more of their tonal qualities. This is the purpose of signal processing equipment. A full service audio post-production facility will probably have many, if not all, of these devices installed near the mixing console:

• *Compressor/Limiter.* The main use is to compress the audio input signal's dynamic range, in terms of volume, when the source material exceeds what the mixing board can be manually set to handle. This is very effective when trying to mix dialogue tracks in which one or more characters speak with a wide variance in volume.

• *Expanders.* The main use of an expander is to increase the dynamic range, in terms of volume, by setting a threshold level to trigger the device when the source audio falls below the threshold setting.

• *Noise Gates.* These devices are used to eliminate very low background noises by setting a threshold level to shut out any audio levels with a volume level below the setting (for instance, using it to clean up dialogue tracks with low volume stage rumble).

• *Room Simulator.* The room simulator is used to generate one of a wide variety of selectable room ambience textures to fill the dry quality of noise gated tracks and to blend tracks that have different ambient qualities.

Figure 8.4: The Boss™ Audio Edit Controller. Courtesy Alpha Audio Automation Systems.

Figure 8.5: The CMX Computer-Aided Sound System (CASS-1). Courtesy CMX Corp.

• *Reverb and Echo.* These effects are used for any instance in which a reflected or delayed sound is useful. One typical example would be to add a slight amount of reverb to dry announce booth recordings in order to give them a more natural room sound.

• *Noise Reduction.* This is an encoding and decoding device that helps reduce the inherent noise that is picked up by transferring through multiple tape generations. It is extremely important to remember that each generation of recording must be encoded during the recording and decoded during each playback for it to be effective.

• *Aural Exciters.* These devices add back the high frequencies and upper harmonics of the audio signal that were lost during the rerecording process. The effect is an increase in the dynamic response of music and voices—in effect, they enhance the audio.

• *Parametric Equalizers.* Parametric equalizers provide a continuously variable control over the boost or attenuation of the audio signal. They select the frequency at which this occurs and vary the bandwidth about which it occurs. Two common uses are to create a narrow frequency ''notch'' to attenuate or eliminate certain continuous frequency background sounds and to ''roll off'' the high frequencies to eliminate high frequency noise or distortion.

• *Graphic Equalizers.* Graphic equalizers provide boost or attenuation at certain fixed frequencies, usually in octave band increments. A typical use is in audio loudspeaker systems to improve the mixing room equalization characteristics.

• *High Pass Filters.* These devices pass any frequencies above the selected frequency setting and block those frequencies below.

• *Low Pass Filters.* These devices pass any frequencies below the selected frequency setting and block those frequencies above.

• *Bandpass Filters.* These filters contain selectable and adjustable high and low frequency threshold settings to pass the frequency range in between the two settings.

• *Notch Filters.* Notch filters contain selectable and adjustable high and low frequency threshold settings to eliminate the frequency range in between the two settings.

• *De-esser.* This is a filtering device used to control the annoying sibilant sounds around 3.2 kHz such as ''S,'' ''Z,'' ''CH'' and ''SH.''

• *Pitch Correction.* This device is used to change the pitch (frequency) of the audio signal to correct minor off-pitch problems, create special mixing effects such as harmonizing and to maintain the correct pitch during program time compression operations (see Chapter 4, using flying spot scanner projection systems).

• *Stereo Synthesizer.* The stereo synthesizer creates an adjustable, synthesized stereo output from a monaural input.

Even though many audio equipment manufacturers make multi-use equipment which feature two or more companion processing devices in a single chassis, most experienced audio mixing personnel know how to use these individual devices for more than one purpose. For instance, the expander can also be used as a noise gate, and the parametric equalizer can be used as a notch filter and a bandpass filter.

Loudspeaker Systems

The function of the loudspeaker system is to convert electrical energy (the audio signal) into mechanical energy (by moving the speaker voice coils), which then becomes acoustical energy (the air movement created by the moving voice coils). Acoustical energy is interpreted by the human auditory system. The reliability and efficiency with which speakers accomplish this task are the basis for selecting a loudspeaker system, and indeed one of the points for evaluating the facility where you are considering taking your project for audio post-production. After all, you want to wind up with what you perceive you are hearing.

I think most people are aware that a single loudspeaker cannot reliably reproduce all of the frequencies in the range of human hearing. Because any large loudspeaker capable of accurately reproducing the low end bass frequencies will not be as responsive in reproducing the high end treble frequencies, there are loudspeaker systems. The loudspeaker systems work on the principle of frequency separation; that is,

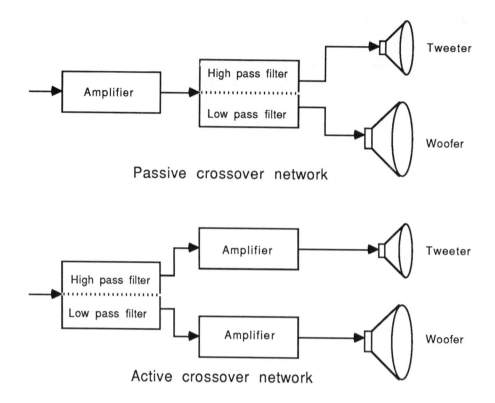

Figure 8.6: Passive and active crossover networks used in studio loudspeaker systems.

frequency ranges are established to drive each speaker in the system. The filtering devices used for separating the frequencies are known as crossover networks. (See Figure 8.6.) These crossover networks are referred to as being either passive or active, or the system can use a combination of each.

Crossover networks are referred to as being passive when they are located after the output power amplifier. This configuration has certain inherent disadvantages because the lack of crossover isolation causes the power levels to be attenuated at the crossover frequencies. In addition, there is also more of a tendency for intermodulation and upper harmonic distortion. However, you only need one power amp for each system.

The crossover network is defined as being active when it precedes the power amplifier. Therefore, it requires at least one power amplifier to drive the woofer section and an additional one to drive the tweeter section; this is known as biamplification. The active network has certain inherent advantages over the passive network because the frequency range isolation helps reduce the possibility of intermodulation and

harmonic distortion. Additionally, it will not transfer intermodulation and harmonic distortion from one section to another.

A simple two-way loudspeaker system consists of one speaker for the low and lower middle frequencies (called a woofer), and one speaker for the upper middle to high frequencies (called the tweeter). The frequency at which the signal is divided (the crossover frequency) usually occurs somewhere between 500 Hz and 1.5 kHz, depending on the physical size of the loudspeaker system.

The three-way system leaves the bass range intact, but divides the treble range into two sections: the mid-range and high frequency. The typical crossover points in the three-way configuration are at about 500 Hz from bass to lower treble, and about 6 kHz from the lower treble to the upper treble. Again, the actual crossover points depend on the physical size of the loudspeaker system.

The four-way system divides the bass frequency range as well as the treble frequency range, with the lower to upper bass crossover at about 300 Hz; the

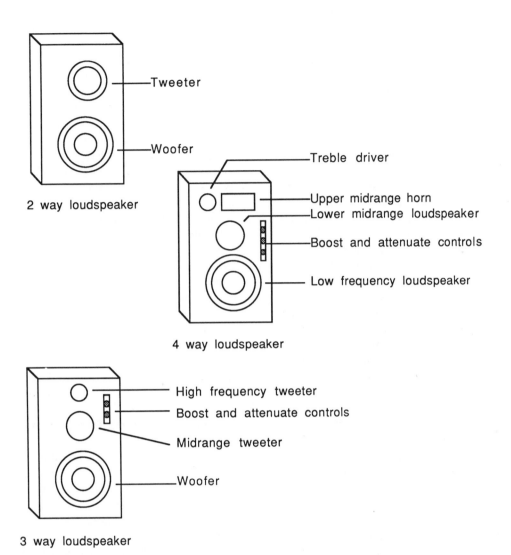

Figure 8.7: Two, three and four-way loudspeakers.

upper bass to lower treble at about 500 Hz; and the lower to upper treble at about 6 kHz. (Figure 8.7 illustrates 2, 3 and 4-way loudspeakers.)

The positioning of the speaker systems in the mixing room makes a tremendous difference in how the system will sound. In most professional mixing rooms the systems directly face the mixing console and are either mounted flush against the wall or, in the majority of cases, mounted in the wall. Not only is the sound level higher (and therefore more efficient) because it doesn't radiate in a full circle, but unwanted sound reflections due to the speaker case to wall distances are eliminated.

Color Video Monitors

Color video monitors are used to view the playback of the videotape recorder in-sync with the audiotape recorder during the ADR, foley, narration, sound-effects editing and final mixdown stages of audio post-production. The size of the monitor is determined by the physical space available, the facility's budget and personal preference. However, many audio post-production facilities do their viewing on large screen projection TVs which provide greater accuracy for inserting in-sync dialogue, music or sound effects, and maintain a much better psychological relationship between the audio and video. The practice of mixing audio with a very large, sophisticated loudspeaker system while viewing on a small TV screen psychologically diminishes the intended visual impact of the program and could result in improper mixing decisions. This would be as equally disproportionate as editing a large screen video production while listening to small, underpowered speakers.

SOURCE AUDIO EQUIPMENT

Source audio comes from many different mediums and audio post-production facilities have to be prepared to handle them. Most of today's source audio is still either record or magnetic-tape based, but the picture is rapidly changing with every new innovation in digital audio. Although digital audio systems such as the Synclavier® and Opus (described later in this chapter) are very high priced in today's market, the situation is bound to change as their popularity and the demand increases—much in the same way as digital video effects systems changed the video industry. In addition, most of the companies that market music and sound effects libraries are presently converting their products to digital compact disc because of their superior audio quality over phonograph records and their relatively smaller size, which translates into reduced storage room and shipping charges. Be this as it may, today's audio post-production facilities should still possess many of the following items.

Audiotape Recorders

Audiotape recorders of the reel-to-reel type are the standard elements in any facility working in audio post-production for video. The basic task of the ATR is to record and reproduce electronic audio signals using magnetic tape, and smoothly transport the magnetic tape from one reel to the other while in the play, fast forward or fast rewind modes. Every audio post-production facility will have some assortment of ATRs since most of today's source audio is still reproduced from, and recorded onto, reel-to-reel ATRs.

There are many aspects to consider when determining which ATRs are suitable for audio post-production work. Price is certainly a determining factor, as is compatibility of the ATR to those used in production and at other facilities. For instance, time code recorded on a Nagra machine will not read when played on an Otari machine. The track capacity is also a major concern since ATRs are available with 1, 2, 4, 8, 16, 24 or 32 tracks using a variety of magnetic tape widths.

Tracks tend to get used up quickly during complicated audio mixes, so to avoid the necessity of doing unnecessary premixes, I recommend that the primary multitrack should probably have at least 16 tracks. In addition, most facilities will also have a 1/2-inch 4 track, a 1/4-inch 2 track and other configurations that they frequently need. (See Figures 8.8 and 8.9.)

The ability to synchronize the ATR is of paramount importance so, given today's time code post-production methods, the ATR must be capable of being frame accurately synchronized and controlled either directly or by installing an interface unit.

Selective synchronization (sel sync) is another feature that converts the record head into a play head. This eliminates the record to play head distance time delay when you need to play back previously recorded material from one track while recording new in-sync material on another.

The ability to perform silent or gapless recordings is yet another consideration. This is especially true when you wish to do in-sync punch-ins during final mixing or when the ATR is being used with a controller capable of 1/100th of a frame accurate editing. The normal punch in–punch out characteristics of many ATRs leave holes or pops at the edit points.

Magnetic Film Recorder/Reproducers

Basically, mag film recorder/reproducers are sprocket driven, single or multiple track audio recorder/players using magnetic film stock. Magnetic film stock is the same size as regular 16mm or 35mm film stock with identical sprocket hole alignment; the difference is that it has a magnetic oxide coating and is used strictly for transferring film sound. Mag film is entirely coated with oxide (full coat) or has one or more oxide stripes running the length of the roll. These are referred to as single stripe, two stripe, three stripe, etc.

Before the introduction of time code production audio recorders, all of the film production 1/4-inch audiotape had to be transferred to magnetic film and synchronized for editing and remixing. Audio for video post-production facilities that needed this ability were required to have one or more magnetic film recorder/reproducers in their equipment inventory.

Cartridge Recorder/Players

The change from using all open reel ATRs began in 1958 with the introduction of the cartridge tape recorder. The cartridge recorder, also called a cart machine, eliminated the need for threading and un-

Figure 8.8: The Fostex E Series 16-Track 1/2-Inch Multitrack Audio Recorder. Courtesy Fostex Corporation of America.

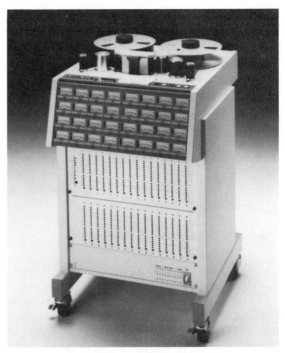

Figure 8.9: The Otari MX-80 24 or 32-Track 2-Inch Multitrack Audio Recorder. Courtesy Otari Corporation.

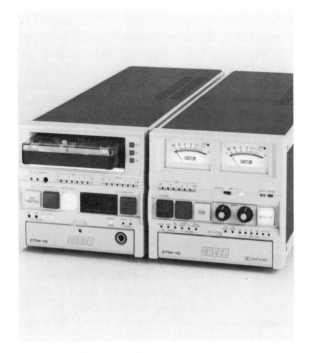

Figure 8.10: The Otari CTM-10 is a mono or stereo time code controllable cartridge recorder and reproducer. Courtesy Otari Corporation.

Figure 8.11: The Sony CDK-006 Auto Disc Loader can rapidly locate, cue and play up to 60 compact discs. Courtesy Sony Communications Products Company.

threading the tape and simplified the cueing and playback of short duration program segments, commercial spots and sound effects. The tape is a continuous loop of standard 1/4-inch audiotape encased in a plastic cartridge running at either 3.75 ips (95.3 mm/sec), 7.5 ips (190.6 mm/sec) or 15 ips (381.2 mm/sec).

The cartridge format is also very useful because several cart players can be stacked and triggered individually or simultaneously by recording a 1 kHz primary cue tone at the head of the selected audio piece, a 150 Hz secondary cue tone to start another cart as the previous one is playing, or an 8 kHz tertiary cue tone for activating yet another similar function. In fact, the new cart players such as the Otari CTM-10 series, shown in Figure 8.10, can also read, cue and synchronize to SMPTE/EBU time code.

Cassette Tape Recorder/Players

The familiar cassette tape recorder/player format is extremely convenient in that it can record up to three hours of material on a small 2.5-inch (6.5 cm) by 4-inch (10 cm) reel-to-reel tape encased in a plastic container. The cassette format became extremely popular in consumer circles over the continuous loop cartridge format because the reel-to-reel format allowed more tape in a smaller case plus two direction playback which doubled the recording capacity. The narrow tape width and slower transport speed made the format very small, lightweight and extremely portable.

The cassette format, however, has limited use in professional audio post-production applications because it is not accurate for cueing and editing. It also has a relatively inferior recording quality due to the tape size and slow transport speed. The recent introduction of recording tape oxides such as chromium dioxide, cobalt treated and metal-particle tape has improved cassette recording quality, but the cueing problem persists. This is why most cassette recordings are transferred to open reel-to-reel ATRs or carts for more accurate control.

Turntables

Professional turntables are available in a variety of models, each one consisting of a drive system, stylus, cartridge and tone arm assembly. There are usually three available speed selections: 33 rpm, neutral and 45 rpm.

The drive systems are engineered to be either direct drive (where the turntable plate is in direct contact with the multi-pole motor shaft) or rim drive (where the motor is attached to a shaft that engages a rubber pressure roller which begins spinning against the side, or rim, of the turntable plate). A third method is the belt drive which is the most inefficient because of its tendency to slip and run off-speed as the belt stretches.

The stylus assembly consists of a metal strip into which a piece of hardened material (usually a diamond) has been mounted. The diamond tip is usually cut into either a spherical or elliptical shape. Spherical tips are the most common because they are durable and produce less groove error. Elliptical tips, on the other hand, actually produce slightly better sound quality, but they also increase groove wear and may dig into the grooves when a record is being back cued.

The stylus is connected to the cartridge. The cartridge acts as a transducer to accurately convert the stylus vibrations to an electrical signal. It also maintains stereo channel separation and keeps the stylus in constant contact with the record grooves.

The cartridge is mounted on the tone arm assembly. The tone arm assembly is designed to smoothly carry the stylus and cartridge laterally across the face of the record as the turntable spins. It also contains counterweights to balance the physical pressure of the stylus on the record.

Compact Disc Players

One of the most recent editions to the audio post-production facility is the compact disc (CD) player. The significant improvement in audio quality and physical compactness for storage provided by the compact disc (and CD player) make it a real candidate to replace today's analog record turntables. The compact disc is digitally encoded on one side by a laser beam which makes it extremely resistant to nicks and scratches because nothing ever physically touches the disc face.

The discs can be loaded individually into the CD player and programmed to instantly cue either forward or backward to play out of sequence cuts with no noise, wow, or skipping while maintaining perfect tracking. New auto disc loaders such as Sony Corp.'s CDK-006, shown in Figure 8.11, can house up to 60 CDs per storage tray with a select-disc access of 16 seconds. Auto disc loaders can be controlled with personal computers using software such as ''The Sound Manager'' available from Pristine Systems, Inc. of Hollywood, CA.

Programs like ''The Sound Manager'' create an inventory listing of all music and sound effects that include descriptions, durations and locations on each CD. Calling up an effect on the personal computer commands the auto disc loader to find the proper CD, load it and cue to the effect.

UNDERSTANDING ADR, FOLEY AND NARRATION STUDIOS

The same equipment configuration can be used for ADR, foley and narration recording; consequently many facilities use a combination recording stage. The difference between ADR and foley is that ADR is in-sync dialogue whereas foley is in-sync sound effects. Narration recording is simply accomplished by recording wild (just reading copy with no video reference), or recording copy to a video playback. In any case, narration is regarded as non-sync dialogue.

The studio such as the one represented in Figure 8.12 is made up of a small control room that is attached to a recording stage and separated by an insulated wall with a viewing window. The control room of the multi-purpose studio contains a multitrack ATR for recording audio, a VCR for video playback, a time-code based controller and synchronizer for accurate automatic replay and record, a small mixing console and a video monitor. It may be more practical for small audio post-production facilities to use their main mixing room as the control room to avoid the cost of buying duplicate equipment.

The stage itself requires one or more microphones, one or more headphone sets and a video monitor. The floor of the foley stage is usually constructed of various hard, soft and crunchy materials for foley walking effects and may be covered with removable carpeting when the stage is being used for dialogue recording.

If the stage is used strictly for foley work there will probably be every sort of gadget and material imaginable stored in the stage. However, if the stage is indeed multi-purpose, then a separate storage room may well be in order.

I advise providing separate microphones and headphones for each person doing narration or ADR looping since dialogue tracks should be kept as clean and isolated as possible.

CHARTING THE AUDIO POST-PRODUCTION PROCESS

There is a logical process to audio post-production just as there is to video post-production. The flowchart in Figure 8.13 traces the path of the typical audio post-production process from the initial laydown stage

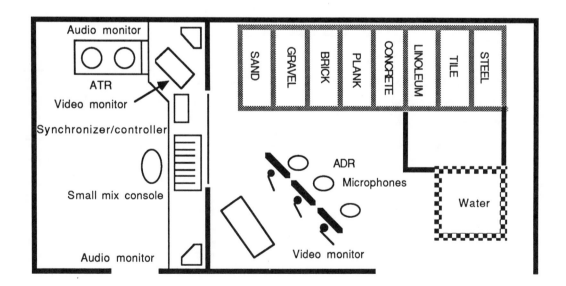

Figure 8.12: Typical ADR, foley and narration stage.

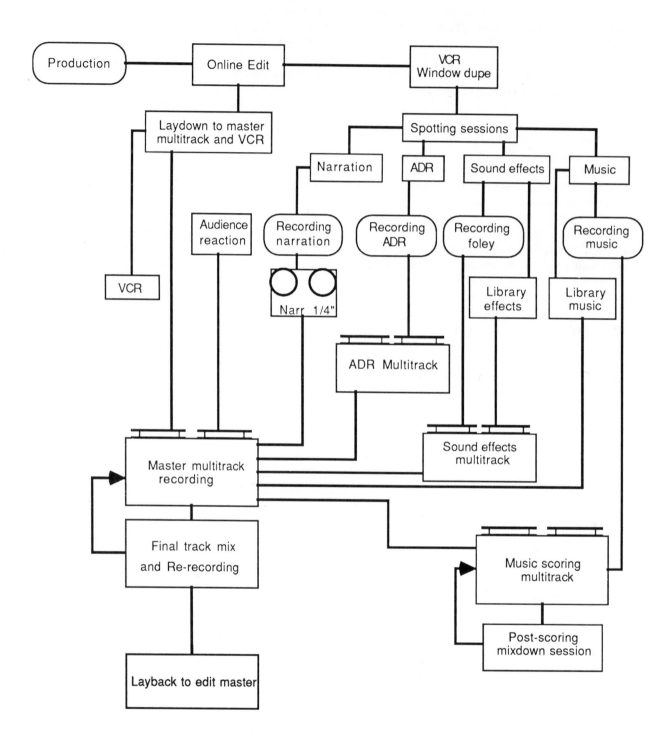

Figure 8.13: Flowchart of a typical audio post-production process.

of recording the edited audio tracks onto a multitrack ATR, to the final layback stage when the balanced tracks are combined and rerecorded onto the edit master videotape. The flowchart also describes the intermediate stages of prelay during which spotting, acquiring, recording and editing are accomplished and the mixing and rerecording stage during which the tracks are mixed and rerecorded.

As we shall discuss later, there are some professionals who prefer to perform these steps before the online edit and still others prefer to complete the audio track even before the production shoot. However, for our purposes, we will assume that the project has already been shot and online edited.

THE LAYDOWN

During the laydown process, the edited audio tracks and time code from the edit master videotape are recorded onto a multitrack ATR, while a simultaneous recording is made on a CMX-style cassette containing time code (either on the address track or one of the cassette audio tracks) and identical time code numbers

are inserted into the video image. The edited audio tracks will most likely consist of alternating production dialogue tracks, dialogue and effects tracks, or dialogue and scratch music tracks. (See Figure 8.14.)

There are two methods that can be used for the laydown. The first method is to record the audio tracks onto precoded multitrack tape stock by using a synchronizer/controller while simultaneously making the CMX-style videocassette. The only advantage of this method is that you have clean time code on the multitrack. The disadvantages are that the tape stock acquires an extra pass of wear during the precoding process and, even though the time code numbers may read the same, there is no guarantee of the exact edit master to multitrack time code phase relationship.

The second and more recommended method is to full record onto new, blank multitrack tape stock. This provides the advantages of less initial tape wear, absolute time code phase synchronization and clean time code because the edit master time code should be passed through a time code regenerator as it is recorded onto the multitrack.

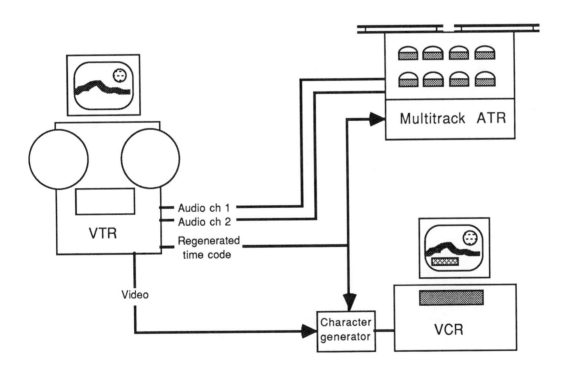

Figure 8.14: Diagram of the laydown process.

THE SPOTTING SESSION

The purpose of the spotting session is to locate and log the time code numbers for building dialogue replacement, sound effects, music cues and narration tracks that are to be added to the final edited master audio track. In order to do this, a 3/4-inch or 1/2-inch window dupe of the edit master is made with the exact SMPTE/EBU time code numbers inserted into the video in large easy-to-read characters. Precautions should be taken to ensure that the character numbers do not obscure any relevant action in the video where the in-sync sound effects or dialogue need to be logged.

After the spotting dupe has been made, the next step is to arrange for the spotting session. If the project is a large scale production requiring a tremendous amount of music and effects, then the session will probably require a composer, along with sound effects, dialogue and music editors. Because of the room and facilities cost for spotting, most producers prefer to view the 3/4-inch or 1/2-inch window dupe with the composer in their office. Music spotting at the normal spotting session is usually only to determine the placement of the music in the video.

Conversely, if the project is a smaller budget non-broadcast production, then the jack-of-all trades theory may well apply. The scheduling will be made much easier, but the task will be lonelier and far more demanding. In the majority of cases, it will be necessary to schedule more than one spotting session for large scale productions. Attempting to schedule a common time when all of the necessary personnel are available is a trick in itself, but efficiency is a consideration. Making the composer, music editor and dialogue editor sit around while the placement of sound effects throughout a program is discussed is a gross misuse of their time.

Methods for Spotting

Methods for spotting vary according to the needs and preferences of the producers. One method is to start at the top of the show and play it through, pausing at each point that needs some help whether it be music, dialogue or sound effects. This alleviates part of the time efficiency problem because everyone becomes sporadically involved. Another method is to start at the top of the program and do a complete spotting for music, then recue and do another complete play for sound effects, and so on. If this is your preference, I

would definitely recommend scheduling separate spotting sessions. In either case, it can be helpful to give extra copies of the window dupe to the personnel involved prior to the spotting session, along with any preconceived ideas and notes, so they will be prepared and the session will go more quickly. No matter which video format is used to record the window dupe, I strongly recommend that it be viewed on a VCR with frame by frame advance, reverse scan and still frame.

Spotting critical cues for sync, sound effects, dialogue replacement and music editing must be frame accurate both for start marks and overall durations. On-the-fly spotting will waste valuable time and money during the sound effects editing portion of audio post-production.

Spotting Cues and Acquiring the Material

The actual spotting process will probably require more than one type of spotting cue sheet because it is more efficient to break down the cues into their respective categories. I have found that acquiring a clean copy of the project's script and noting rough cue marks for known effects requirements during video editing helps in audio post-production because the notations can be seen alongside the familiar scene dialogue or action descriptions. Complicated projects will probably require track assignments, or individual multitrack reels, for some or all of the effects categories. If the person making out the spotting sheets is unfamiliar with assigning tracks it may be necessary to have the audio mixing person at each spotting session. It may also be helpful to consult with the mixing person in advance of the session so the track assignments can be properly planned and the required equipment made available.

Sound Effects

Sound effects may consist of generic door closings, doorbells, telephones or any other familiar day-to-day sound. Some scenes may even require distinctive location ambience such as one would encounter in a factory, airport or hospital.

The spotting session will provide the start points and durations in the edit master for including these effects. However, the person responsible for editing these effects should have a copy of the window dupe for the areas where the complicated effects sequences take place. For instance, editing airport location ambience throughout a scene will require as much plan-

Program title						Episode	
Page of			Date			Engineer	
Page	Effect	Source	Time in	Time out	Trk	Comments	

Figure 8.15: Sound effects cue sheet.

ning and work as scoring a scene. Adding sound effects like jet engines, passenger hubbub and PA announcements requires careful attention. The proper mood should be set without affecting the actors' dialogue. A scene such as this may well require its own multitrack recording and mixdown session.

A sound effects cue sheet, such as the one in Figure 8.15, provides the sound effects editor with necessary information. The heading should contain the basic information to identify the project, the date of the project and who was responsible for spotting the cues. The rest of the sheet is divided into seven columns:

Column 1—indicates on which page of the script a specific effect is needed.

Column 2—identifies the actual sound effect type (i.e., a car crash, rainstorm, siren).

Column 3—indicates where the sound effect can be located. This column could be left blank during spotting and filled in by the sound effects editor as he or she locates the sound effect in the sound effects library. This column can also be used to indicate effects that have to be created.

Column 4—is the precise edit master start time in SMPTE/EBU time code.

Column 5—is the precise edit master stop time in SMPTE/EBU time code.

Column 6—indicates on which multitrack channel the effect will be assigned.

Column 7—is for entering notes such as how the effect should start, the effect duration, how the effect should end, or any other appropriate comments.

The spotting session should not only help determine which effects will be needed and where they will go, but it should also give the necessary information about where the effects can be found. Sound effects can be acquired by using prerecorded effects, by wild recording of the actual effects, or by recording simulated effects on the sound effects stage.

Library Sound Effects

Prerecorded sound effects libraries that provide hundreds of effects for one overall charge, are readily available for purchase. These libraries are available on audiotape, record albums and, most recently, compact discs. I highly recommend examining each prospective library for its variety of available effects before purchase. The ambience of each effect makes a great difference in how well it will work. For instance, exterior gunshots will sound out of place when used in an interior setting, as will a hollow sounding door knock when it is placed over a shot of someone knocking on a solid door. Check for the companies offering the library sets that include the effects in several different ambient settings. Of course, each effect can be tailored to some extent during the sound effects editing and final mixing steps, but why waste unnecessary time and money.

High budget productions can take advantage of the new digital audio systems such as the Synclavier® or Opus (to be discussed in more detail later in this chapter), for recalling, tailoring and sub-frame editing their inventory of effects.

If the project does not warrant the expense of purchasing a sound effects library, then an audio post-production facility that owns a library, knows where to get one, or which can quickly record or create the required sound effects on stage should be contacted. Many non-sync sound effects (also called atmosphere) can easily be recorded or created on the stage, where the room ambience can be more accurately controlled. The materials that generate the sound should be brought onto the stage, recorded separately, and then added to the multitrack during the sound effects editing step. It is inevitable that some effects will have to be created by different means, since it is unlikely that a recording facility will permit the lighting of a fire, or the creation of a car accident, on its stage. Any experienced sound effects editor will know which tricks to use. For example, stepping through wads of old 1/4-inch audiotape creates the effect of walking through grass. Kneading stiff paper and cellophane creates the sound of fire, and sifting through a pile of empty plastic film cores creates the sound of sea shells bumping together. The required sync sound effects, however, may have to be specially created by a process known as foleying.

Foley Sound Effects

Foley is the process by which sync sound effects are replaced or added during audio post-production. The process requires the use of a synchronizer, an

Program title					Episode	
Page of			Date		Engineer	
Page	Time in		Time out		Trk	Description of effects

Figure 8.16: Foley effects cue sheet.

audiotape recorder and a VCR to play back portions of a scene to people on a sound effects stage (called foley walkers). The sound effects concerned could be actual effects that sound false on the production tracks (gunshots are a good example), effects that may have been lost because of improper production miking or location noise, or effects that were under sync dialogue sequences which require automatic dialogue replacement. If the project needs a lot of foley work, I would suggest using a video playback without numbers in the picture. The time code information can still be available by simply installing a time code reader above the control room video monitor.

Figure 8.16 illustrates one example of a foley effects cue sheet. The basic heading contains the relevant information needed to identify the production. The sheet is divided into five columns.

Column 1—indicates the script page number in which the sound effect occurs.

Column 2—is the precise edit master start time in SMPTE/EBU time code.

Column 3—is the precise edit master stop time in SMPTE/EBU time code.

Column 4—indicates onto which track of the foley multitrack the effect should be recorded.

Column 5—is used for detailed descriptions of the effect along with any other helpful remarks.

Automatic Dialogue Replacement

Automatic dialogue replacement (ADR) is a process similar to foley, except that it is done with the actors recreating their lines. The ADR editor plays back the synchronized VCR and ADR multitrack to the actors in a sound booth while recording their in-sync lines at the corresponding video master time code locations. This is done repeatedly until the producer is satisfied with the performances. ADR is the updated version of what filmmakers called "looping." Many years ago, the process of looping required actors to recreate their lines while they watched a repeating loop of film. Their new lines were recorded on magnetic film stock that was interlocked to the film loop projector.

An ADR cue sheet, like the example in Figure 8.17, is formatted to contain the obligatory heading information with seven columns for entering ADR information.

Program title						Episode	
Page of			Date			Engineer	
Page	Character	Time in		Time out	Trk	Scripted dialog	Comments

Figure 8:17: ADR cue sheet.

Column 1—contains the page number in the script where the dialogue occurs.

Column 2—indicates which character is delivering the line.

Column 3—is the precise edit master start time in SMPTE/EBU time code.

Column 4—is the precise edit master stop time in SMPTE/EBU time code.

Column 5—indicates which tracks of the ADR multitrack will contain the dialogue takes.

Column 6—contains the actual scripted dialogue to be replaced.

Column 7—is for entering any notes or relevant remarks concerning each take.

From a technical standpoint, in-sync dialogue replacement is a simple process. But, from a performance standpoint, it is a process best left to professional actors and actresses who are experienced in the area. Producers of projects using nonprofessionals, children, or actors and actresses who are inexperienced with the ADR process, may quickly discover that they should have taken the extra few minutes to do pick-ups during the production shoot.

Audience Reaction Sound Effects

Audience reaction sound effects include laugh tracks such as those heard on television sitcoms, the ''oohs'' and ''aahs'' prevalent throughout syndicated game shows, and the whistling and applause usually heard on variety shows.

While the use of these effects is a matter of personal taste, my own feeling is that they should only be used to augment those effects created by a ''live'' audience; anything else intrudes on the integrity of the production. Thunderous applause after a remote location musical number, or audience laughter over a humorous moment shot during an exterior street scene makes a mockery of the very performance that is to be enhanced. If these effects are to be mixed into the production track, great care should be taken that the sound is plausible. The ovation of a Kennedy Center audience will seem bizarre if it is placed over a scene shot at a small nightclub or a speech shot at a stockholders' meeting. In fact, most of the comedy shows that I have worked on record the ''live'' audience and give

the recording to a specialist who adds audience reactions. These specialists use keyboard controlled ''laugh'' machines that are plugged into the audio mixing board during the editing or mixing session. The ''live'' audience recording is included in the repertoire of effects so that any smoothing and filling that may be required throughout the program will blend properly.

Narration

The spotting process for locating narration points is usually not as accuracy intensive as spotting sound effects. The biggest factor in this category is the duration of each announce segment. However, if the project is tightly edited so the narration must coincide with certain visual points, a frame accurate spotting list should be completed.

The most efficient method in such cases would be to wild record the narration before video editing, then cut the video to it. This would alleviate the need for spotting narration points altogether.

Music Cues

Music can be one of the most important elements in the audio post-production process. The proper blending of dialogue and music, video cut to music, or the use of music as background scene ambience helps to define the scene for the viewer.

Two of the first considerations in music spotting are how much music to add and where to put it. The intent of the project, whether or not the message stands by itself, and whether or not the music is needed to augment certain passages or underscore the entire program will determine where the music should go.

The most common places for music are at the head and tail of the program; spots in between are open to wide interpretation. There have been shows that I have edited that were never intended to have internal music. I found, however, that certain feelings and moods were not being sufficiently conveyed to the viewer and the subtle inclusion of properly scored music made these scenes come alive emotionally. It is important to remember, however, that the music is intended to enhance the moment, not upstage it. Judicious restraint applied during music spotting is the wisest course. Conversely, do not be afraid to experiment with music in uncertain areas. Nothing is locked in until the final mixdown. If the music doesn't work, eliminate it.

Another major consideration is what style of music to include. Are you adding music to establish the geographic location or historical period of a scene?

Are you trying to appeal to certain age or ethnic groups with your message? Consider the emotional reaction you want to achieve from the viewer by selectively changing the composition and pacing of the music. These are a few of the factors that need careful planning for the music to be effective.

Honky-tonk piano music over the exterior shots of older multistory buildings would convey a 1920s ambience to most viewers, whereas a fast-paced orchestral score with the brass section simulating automobile horns would change the same shot sequence to the hustle and bustle of a large modern city.

Music is a great persuader. The ways in which it can be used to supply age and ethnic group appeal is very evident by the scoring of most fast food, soft drink and designer jean commercials. Modern upbeat rock, rap or bluesy commercial themes persuade the viewer to get with it, be part of it, be in.

The proper music score mixed with effective visuals, can induce every possible emotional level from the viewer. Changing the scoring pace and orchestration during a scene will signal a mood transition for viewers, alert them as to what is coming or make them recall previous events in the program. There are two ways to procure music for production: use library music or compose and record original music.

Library Music

Library music is the simplest and least expensive to use. Music libraries provide finished original music to heighten or underscore almost any possible combination of scene requirements. They are commercially available on either record albums or compact disc. Fees vary according to the medium used. "Blanket" licenses are also available to producers. These provide producers with the unlimited use of the library for a specific time, say one year, with the fee based on the licensee's type of industry. Educational and non-broadcast producers are charged less than broadcast and cable TV producers because of the different music licenses required for air and cable broadcasts, while producers of retail videocassettes have yet a different licensing agreement.

Producers can also get production licenses that give them limited music use. These can be paid for in the form of "per drop" rates (payment made each time music is used from a particular cut) or "production blankets" (graduated payments for using less than 5 minutes, less than 10 minutes, less than 15 minutes, etc.). Remember that any recorded material that is copyrighted may not be used for nonprivate purposes without permission from the holder of those rights. Payment for use of copyrighted material is usually required.

Recording Original Music

By recording original music, the producer obtains the most flexibility for music tailored exactly to the needs of a particular project. Unfortunately, it also provides the producer with the highest cost and the slowest completion turnaround of the two options. Composing and recording original music for a production requires the hiring of a composer and possibly a lyricist. An arranger, music copyist and printers to write the scores for each musician, a musical director and of course the musicians are also needed. In addition, there may be a scoring stage rental fee, salaries for the scoring stage crew, instrument rental costs and instrument hauling charges. Another additional charge of renting video projection equipment would be required if the score was to be done to video playback.

These are just some of the overall considerations for original music scoring; each production situation is different. Composers familiar with synthesizers can also cut down on the cost of scoring if the "electronic sound" is right for the project. In fact, with the introduction of synthesizers (such as the Yamaha Emulator or the Synclavier®) music scoring is being done during the prelay session. This provides the advantage of first generation quality sound, frame accuracy so time is not wasted slipping mixdown tracks, and ultimate flexibility in the mix. Since the music tracks are not locked, different instruments can be enhanced or ducked under if they are interfering with the dialogue.

MULTITRACK RECORDING

The primary objective of multitrack recording is to have all of the audio sources recorded in-sync on one reel of tape (or more if the project is complicated). In doing so, it is important to assign the recording tracks in a manner conducive to easy and efficient mixing. Most track assignments wind up being a matter of mixer's personal preference. There are certain things to consider, however, these include keeping the raw source tracks in one group and the mix tracks in another, isolating the time code from the audio tracks and keeping two high-volume level (hot) tracks separated to prevent cross talk.

Assigning Tracks on the Master Multitrack Recording

Most final mixes end up having separate tracks containing mixed and equalized dialogue, sound effects and music. This type of mix is referred to as a three stripe mix—a term adapted from film terminology. Many other mixes also require an additional track of audience or off-camera narration. These are referred to as four stripe mixes. Bear in mind that this refers to monaural track mixes; stereo mixes would require double the amount of tracks. For example, the three stripe mix becomes six stripe and the four stripe mix becomes eight stripe. Stereo mixes may require smaller audio post-production facilities without 24 track recorder capability to mix over to a second multitrack recorder.

Tamara Johnson, a veteran Hollywood audio mixer, prefers to separate track assignments into three groups: the mix tracks, the raw source tracks and the time code plus guard band tracks. By studying the example in Figure 8.18, we can see that she prefers to assign the tracks so that the final mix section takes up the first eight tracks of the recording and the raw source audio tracks take tracks 9 through 22. The guard band and SMPTE/EBU time code track take up the rest.

If the mix is mono only, she prefers to use only the odd numbered tracks (1,3,5,7) while leaving the even numbered tracks clean. This opens up four additional tracks, but more important, it allows her to preserve the efficient track assignment pattern she is familiar with so time and energy are not wasted dealing with unfamiliar track locations.

While there are eight tracks available for music, remember that stereo cuts the multitrack by half. Each segue will require at least four tracks—two for the outgoing music and two for the incoming music, with the possibility of two additional tracks for bridge material. This problem is discussed later in this chapter.

The four sound effects tracks do not necessarily duplicate in stereo because some static effects can be spatially panned onto the mix tracks (although lateral motion stereo effects such as cars driving by and airplanes landing or taking off may require duplicate tracks). The guard band track can be left blank, but the more common practice is to record the 59.94 Hz TV sync signal as a safety precaution so the tape will still synchronize if something happens to the time code signal on the final track.

None of the preceding track assignments are mandatory except the time code and guard band tracks. Track assignment is left up to the capabilities of the equipment and the personal preference of the audio mixer. However, whichever track assignment method is going to be used, I strongly recommend making a track assignment chart (such as the one in Figure 8.19) so the assignments are clear to anyone concerned. A copy of the chart should also be left in the multitrack reel box.

Figure 8.18: Track assignments on the multitrack recording.

Tracks	Material
1 and 2	Mix sound effects
3 and 4	Mix dialogue
5 and 6	Mix music
7 and 8	Mix audience or narration
9 through 16	Source music
17 through 20	Source sound effects
21 and 22	Edit master audio
23	Blank or 59.94 Hz
24	SMPTE/EBU time code

TITLE _____			
EPISODE _____			
REEL _____ OF_____			
ENGINEER _____ DATE ____			

1	2	3	4
5	6	7	8
9	10	11	12
13	14	15	16
17	18	19	20
21	22	23	24

Figure 8.19: Typical multitrack assignment chart.

Assembling the Dialogue Tracks

Dialogue tracks are normally assembled in one of three ways:

• Completely rebuild all of the production dialogue tracks from the original production tapes.

• Pre-build and mix the final audio track before the online edit.

• Use the existing audio tracks from the edited master videotape.

As I mentioned earlier, there are several network film productions utilizing electronic post-production that prefer to treat the video edit master dialogue tracks as merely scratch tracks to be replaced in audio post-production. Since the original production shoot uses 1/4-inch Nagra recorders which also record time code at the time of shooting, the time code will follow whenever the film and production audiotapes are transferred in-sync to videotape. Consequently, the person doing the audio mixing can use the online edit decision list to accurately reassemble the production 1/4-inch onto the master multitrack tape. This is done primarily to obtain the flexibility of using multiple tracks for dialogue equalization and mixing. It also saves two generations of audio rerecording (one during the original transfer and one during the online edit).

The method of pre-building and mixing the final audio track before online editing is used by many film commercial producers to achieve a final, approved audio track before the online edit session. They can do a standard dialogue assembly and mixing session while viewing workprint copies of their commercials. This can be the most efficient of the methods for commercial work because the approval of the commercial comes from many agency levels. It is more efficient and economical to do several quick temporary mixes, get the final approval and do the final mix, than to repeatedly re-edit with expensive online sessions.

Many commercial producers are still more comfortable doing a standard film-type mix using edited mag stripe film (plus they already own the necessary editing equipment). However, the introduction of production audio recording with time code may convince more of them to change to electronic post-production methods in the future.

Music videos are another category for pre-online audio mixing. Pre-mixing the production tracks gives them the advantage of having time-coded, synchronized music tracks for the production playback. In addition, they make an identical time-coded cassette base containing the synchronized music mix for workprint editing, and a reel of identical time-coded 1-inch videotape base containing the final synchronized music mix which will serve as the video online master recording (see recording digital audio on videotape later in this chapter). The majority of music video producers follow this procedure unless the particular video contains on-camera dialogue not normally present in the music mix. In this case they are more likely to treat the music as a scratch track and do a final post-production remix, thus preserving the integrity of the original mix.

The third method of using the existing edited master audio tracks is the most common and generally most efficient method of dialogue assembly. The tracks may consist of one dialogue track, split dialogue tracks, dialogue and location ambience, dialogue and audience reactions or some other appropriate combination. This method requires that more care must be taken during the online edit to ensure that the audio levels do not fluctuate wildly and distort, and that takes with vastly different ambient or EQ values are not adjacent on the same track. In such cases it is far better to split the tracks and allow some ambience overlap between each edit so the mixer can smoothly segue between the tracks. If more than the customary two tracks are needed for on-camera audio, I suggest that the video editor either make ''B rolls'' for the extra tracks or note on the edit decision list the precise in/out edit points. This will allow the audio mixing person to edit those sequences directly from the production videotape onto the multitrack tape. Of course, this is providing that the audio post-production facility owns and has the ability to control the VTR format used in the production.

Assembling the ADR Tracks

Assembling the ADR tracks involves synchronizing the ADR multitrack and rerecording the tracks onto the master multitrack at the identical time code numbers (although in reality it still may be necessary to slip the ADR tracks slightly for a better lipsync match). If there are sequences with more than one person doing ADR lines (such as group discussions or arguments), then individual tracks, or even individual multitrack reels, will be assigned to each person. Clean room or location ambience should also be recorded on yet another track to cover each of the ADR sequences so no "void" or radical ambience change is evident when they are used in the mix.

Assembling the Narration Tracks

Selected wild narration tracks are rerecorded onto one or more tracks of the master multitrack at the time-code points listed on the cue sheet. For instance, individual tag lines for regional commercials could be recorded at the same time-code location, but on different tracks, so that several versions of the same spot could easily be laid back. Shorter narration sequences can usually be rerecorded intact, but long sequences may need retiming and editing (unless they were recorded to video playback). This should not present a big problem, however, if the narration tape has time code, since any editing can be done with a controller just as in video editing. A common practice is to record the wild narration using a time code 1/4-inch ATR, then transfer it to 3/4-inch videotape for offline editing. During the prelay session, the original 1/4-inch audio can be reassembled on the multitrack, thus maintaining the highest quality of audio. If there is no time code on the tape, it may be easier to physically splice the narration tape, especially if various portions of several takes are to be used in each complete passage.

I would like to point out again that the background ambience or music must provide some fill in order to avoid the tracks sounding dry and empty. The additional step of adding reverb to the narration will also give it more life.

MIXING DOWN MULTITRACK MUSIC

I have included this section not with the intention of trying to provide an in-depth discussion of the intricacies of rerecording mixdowns, but merely to indicate the procedural steps in the mixdown process. Let's assume for the sake of discussion that we are in the main mixing room, the equipment is cleaned, calibrated and set for the proper audio playback and recording levels. We have made sure that the monitor speaker feeds from the mixing console are set so one channel is delegated or panned to the extreme right, while the other channel is delegated or panned to the extreme left. Since the mix will probably need to be mono compatible, the console monitor switches should also be set so it is easy to switch back and forth between stereo separation and monaural.

Each track of the unmixed multitrack is fed to an input fader position on the mixing console (so it's obvious that the mixing console must have at least as many input fader positions as there are music tracks). The input equalization pots for each channel need to be adjusted so that the replay of the multifrequency test signals from each individual track will be reproduced at a constant, uniform level (O VU). The next step is to assign the tracks as to left, right or panned somewhere in the middle (providing that this is a stereo mix to begin with).

The first run-through is to check and adjust the levels and spatial positioning of the vocal tracks (if there are any). Otherwise, the levels, blending and spatial positioning of the various instrument tracks are checked and readjusted accordingly. This step can be a time-consuming, repetitive process of listening and adjusting.

Individual tracks are then re-equalized, if necessary, and any special effects such as digital delay, flanging or other sound-altering devices are added and adjusted. Generally, the rhythm tracks are dealt with first, followed by other instrumental tracks in any order of preference. Vocal tracks are handled last. It would be chaos to try listening to and adjusting all of the tracks simultaneously.

The entire mix is listened to for final balancing, blending, spatial positioning and mono compatibility. When everyone is sufficiently satisfied with the mix, it is rerecorded onto a new multitrack tape. First, a series of multifrequency test tones are recorded at the head end of each audio track containing the mixdown. This could simply consist of one track for a mono mixdown, two tracks for a stereo mixdown, three tracks for a three stripe mixdown of the separate rhythm, string and horn sections, or more than three tracks depending on any additional mixdowns that are required. For instance, if a live stereo concert is to be conformed to the edit master videotape, the balance, mix and spatial relationships of the audience and any stage dialogue must be maintained throughout the final mix.

The music mixdown and any additional mixes are rerecorded onto the new multitrack tape, while matching time code is simultaneously recorded onto the track at the extreme edge from the mix tracks and 59.94 Hz sync pulses on the track adjacent to the time code.

MUSIC TRACK CONFORMATION

Music track conformation is the process of rerecording the final music mixdown tracks onto the master multitrack reel in exact sync to the scratch track music.

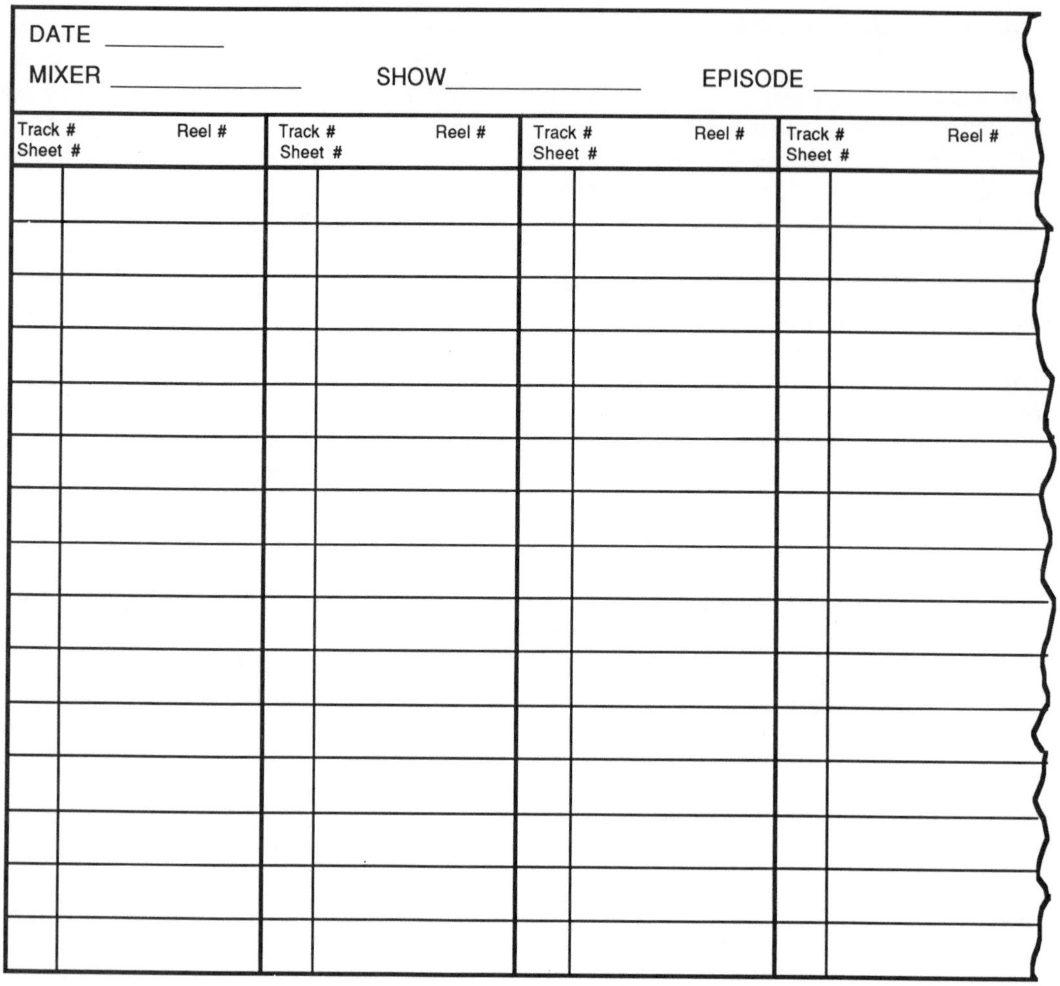

DATE _____							
MIXER _____ SHOW_____ EPISODE _____							
Track # Sheet #	Reel #	Track # Sheet #	Reel #	Track # Sheet #	Reel #	Track # Sheet #	Reel #

Figure 8.20: The mixer's rerecording cue sheet.

This is done in much the same way as the dialogue tracks are assembled. The music mixdown reel should contain the same time-code numbers as the unmixed scratch track recording to avoid the time-consuming need to calculate number offsets. Music conformation can be, and in some areas probably still is, done without time-code reference control. This is a slow, more tedious process requiring the second audio person to wear headphones and listen for the telltale echo that indicates an out-of-sync playback condition between the scratch and mixdown tracks. Internal music edits made during the video online session are also extremely difficult to satisfactorily reproduce without

the accuracy of time-code based controllers.

The basic procedure will require at least two multi-track ATRs, the mixing console, a time-code based synchronizer/controller and possibly video playback. First, the multifrequency test tones from each of the mixdown audio channels are played back through the individually assigned input fader positions on the console, while each channel equalizer is adjusted for a constant uniform playback level (O VU). The controls will normally be left in these positions throughout the conformation since this is the final accepted mixdown condition. Remember that different mixing rooms may sound slightly different to a trained ear. Constant readjustment of the channel equalizers to balance the apparent difference is analogous to readjustment of the playback hue during video online editing every time it is viewed on a different color monitor and will only result in hurting the quality of the end product.

Next, each relevant audio edit must be re-performed on the preassigned track (or tracks) of the master multitrack according to the time-code edits specified on the online edit decision list. This includes the music mixdown and any additional tracks that needed remixing or balancing.

Two reminders for a successful conformation are:

1. Each mixdown track must be put through a compatible decoder unit if any noise reduction encoding was used in the remix. In addition, if noise reduction is to be carried on throughout the process, each track must be put through another encoder unit before rerecording it onto the master multitrack.

2. Stereo tracks must be checked to ensure that they are being reproduced and rerecorded in phase both with each other and the other existing tracks, otherwise frequencies simultaneously appearing in each channel will cancel out.

PREPARING THE RERECORDING CUE SHEET

The importance of the audio mixer's rerecording cue sheet is analogous to the video editor's online edit decision list. It tells the mixer where everything is located on the master multitrack reel, which is particularly important if the mixer was not the same person doing the prelay process. Although cue sheet formats will vary from facility to facility, they convey the same necessary information.

The cue sheet, example shown in Figure 8.20, is used by many audio post-production facilities. It is set up in vertical columns with each column representing one track of the master multitrack reel. Each line in the column represents a discrete amount of elapsed time which will change depending on the overall running time of the project and the complexity of the mix. One or many pages of the cue sheet may be required for an entire project.

The first category in the column indicates at what SMPTE time code point each cue happens. The second category is used to briefly describe each cue. I recommend entering a brief, concise cue description because the column space is limited and the mixer should not have to guess its meaning. Indicator marks should

also be entered to tell the mixer when and how the cue starts, how long the cue lasts, and when and how the cue stops. For instance, a vertical line drawn on both sides of the cue description will indicate where it starts, how long it runs and where it stops. If the cue is to be faded up, faded out or crossfaded we might draw arrows (◄—►) at each end of the vertical lines. The actual form of the indicator marks is up to the preference of the mixing team, just as long as they are legible and everyone understands their meaning.

Figure 8.21 is a simplified example of a rerecording sheet for dialogue and music cues. We can see that the dialogue editor has A/B staggered the two tracks and used elapsed minutes and seconds (but no frames) to indicate where the edited dialogue starts and stops. The actual dialogue has been indicated by notating the characters (Eric is on track 9 and Karl is on track 10). A portion of their dialogue for each cue entry has also been notated. The mono music cues are also A/B staggered on two tracks (stereo mixing will require two additional tracks for the other channel) which include the minutes, seconds and frames time code for each cue start and stop, plus the music cue identification numbers in sequence.

MIXING AND RERECORDING

The prelay process has taken care of compiling, synchronizing and editing all of the material onto the master multitrack reel. Now we are ready for the mixing and rerecording process. Mixing and rerecording is the step by which the dialogue, music, sound effects, narration and audience reaction tracks are equalized, balanced and adjusted to each other. They are then rerecorded back onto the preassigned tracks of the master multitrack reel . At this point all of the previous effort comes together and any technical problems with the tracks should have been worked out. Neither the mixer nor the client wants to waste time trying to repair problems that should have been handled earlier in the process. As we mentioned earlier in the chapter, the mix process should result in separate tracks called a three or four stripe mix (six or eight for stereo); dialogue, music, effects and narration. In cases where all of these audio elements are being used, audience reactions are normally treated as effects.

Even if the end product is a relatively straightforward monaural track, mixing the elements into three or four separate track groups provides a great deal of flexibility. There are many projects which may ultimately require changes in the dialogue or music tracks, especially when the project is dubbed in a foreign language or uses specific dated data like sales figures, regional projections and limited-use music cues. Separate mix tracks will provide the convenience of changing one source of the overall mix without the necessity of redoing the entire mix.

The best method of mixing is to take each step one at a time instead of trying to mix all of the tracks at once. Unless the mix is extremely simple, it would be a mistake to try smoothing dialogue sequences while simultaneously fading in and out music cues.

Jerry Clemans, another veteran Hollywood audio mixer, prefers to work on the dialogue and voice-over tracks first because most projects are designed to play off the dialogue. He feels that if the dialogue is not understandable the message will be lost. After the dialogue tracks are blended, level adjusted and equalized, his next step is to mix either the music or effects tracks—whichever is dominant. This includes simply fading in and fading out the cues from one track or segueing alternating cues on several tracks. Mixing the tracks in this manner requires several complete passes through the length of the project; at least one for each final track, and one for an overall mix assessment at the end of the session.

Monaural Mixing

Mixing the final monaural track requires monitoring the combined outputs throughout the mix session even

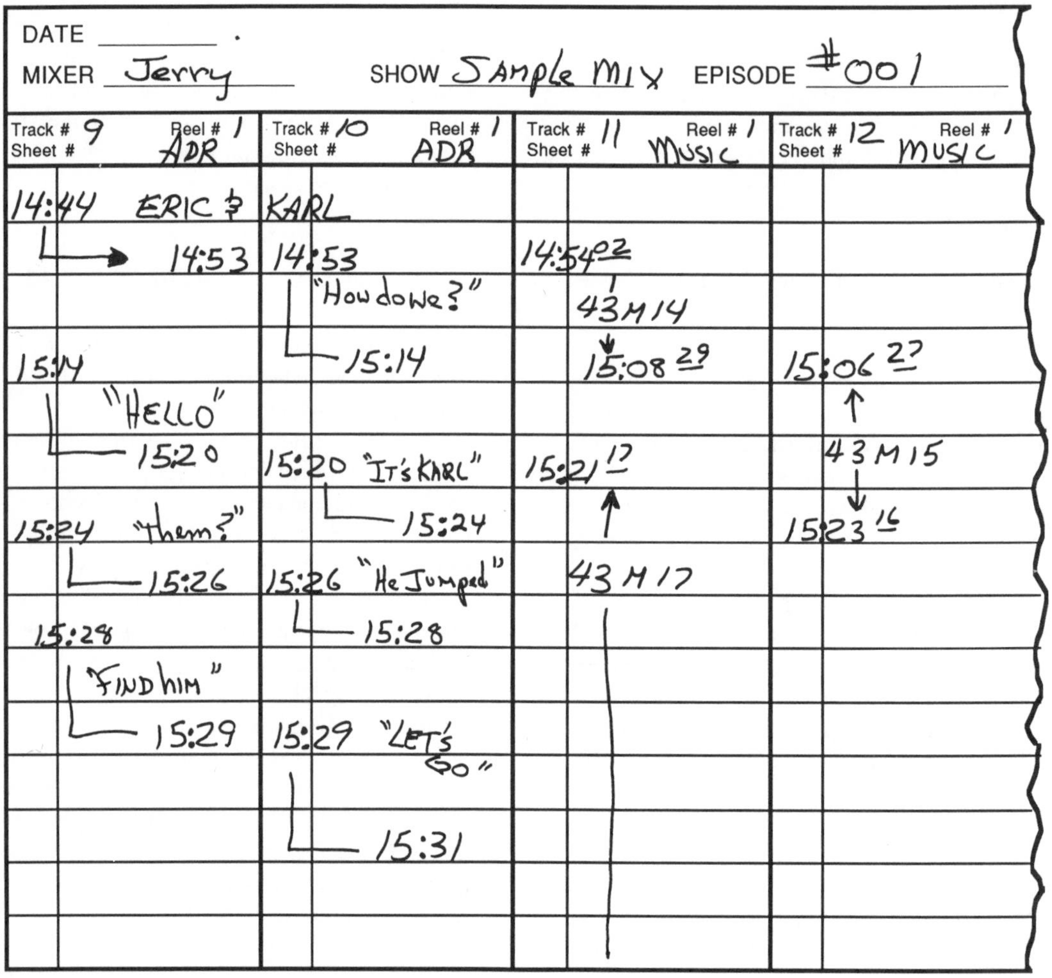

Figure 8.21: Typical marking of a rerecording cue sheet.

though the tracks remain separated. Since all of the mix elements are completed in this manner, the monitored playback of the mixed tracks will sound correct. When the mix is finished, the entire project should be listened to through a speaker system approximating the type of equipment through which the project will ultimately be heard. In effect, projects for videocassette or TV transmission should be listened to through small speakers, whereas large audiovisual projects are more likely to be reproduced through large speaker systems. The desired result is that each of the three or four separate tracks is finished, including any level changes for music segues or scene transitions. This permits a simple, straight across one-to-one mix to the VTR during the rerecording or layback process.

Stereo Mixing

Stereo mixing is different not only because there is twice the amount of tracks involved, but also because of the spatial relationships inherent in the mix. Stereo has the unique quality of adding location and depth to the mix. Consequently, in addition to the mixing parameters in monaural, we have the added concern of deciding in which direction the sound should be coming from. Even though large screen projection TVs are popular, rerecording for television is still basically a small screen medium. Mixing dialogue sequences to the extreme left or right positions does not have the same impact of a wide screen feature because there just is not enough visual separation. As a general rule, the dialogue should almost always be mixed to the center, with the music and effects surrounding it to add depth and fullness to the mix.

Stereo to Mono Compatibility

Mixing stereo tracks that must be compatible with mono is a problem in today's U.S. television industry. The problem is that any signals appearing in both channels will have an additive effect when they are combined. For example, stereo dialogue tracks mixed at the center at O VU will be reproduced at $+3$ VU per channel or $+6$ VU, possibly causing distortion and certainly overwhelming the rest of the mix. Conversely, synthesized music tracks mixed to the extreme edges will also suffer in relation to the rest of the mix because of the stereo to mono phase cancellation effect. Either occurrence severely affects the outcome of the final mix. To combat this effect, many mixers feel that the center tracks should be mixed at -3 VU

lower than normal. In fact, most audio mixers adjust the stereo track levels while listening to a mono monitor feed. Unfortunately, stereo to mono compatibility is a quality compromise at best.

THE LAYBACK

The layback process (see Figure 8.22) is the final step in audio post-production. This is the stage during which the final mixed and approved three or four stripe (six or eight stripe stereo) audio tracks are combined and recorded back onto the edit master videotape in perfect synchronization. The final summed mix can be either monaural, two separate tracks (as in the case of dialogue separated from music and effects for foreign distribution), two track stereo, or even three or four tracks of audio with the newer VTR formats that we discussed in Chapter 2.

It is also at this step that the decision whether or not to add noise reduction encoding to the final audio track is made. Tape hiss reduction when making subsequent copies of the final master is certainly a desirable feature, but it may or may not be a major concern depending on the audio quality and final use. However, there are two things to keep in mind about noise reduction: The master audio can never be replayed properly without using a decoding device, and different noise reduction systems are not compatible. In other words, encoding the edit master audio tracks for Dolby A noise reduction means that each time the master is replayed, the audio tracks must each go through a Dolby A decoder unit in order for the proper sound to be heard or recorded. If encoding and decoding is a problem, then noise reduction should probably not be added.

Upon completion of the layback process, I highly recommend a complete viewing of the finished program to assure that what was supposed to happen, did happen. The layback is a relatively simple process, but faulty switches, audio patch cords, record level controls or any number of other malfunctions have been known to pop up at the most inopportune times. The extra half hour or hour involved in a careful viewing is a cheap enough price to pay to ensure the finality of the project.

DIGITAL AUDIO RECORDING METHODS

Although digital audio recording for video productions (other than music videos) is still in the future for the vast majority of projects, I do not feel that this chapter would be complete without some discussion of digital audio recording methods. Before digital audio recording becomes commonplace in the video industry, the persistent problems of high equipment costs, different record/play head configurations and recording formats will need to be resolved.

Basically, digital audio recording differs from analog audio recording in that the actual recorded signal consists of a series of coded electronic pulses instead of an electronic waveform which resembles the waveform of the original source audio. Digital recording has several advantages over analog recording. For instance, digital recording allows higher dynamic ranges without fear of distortion, no tape noise, no wow and flutter and no deterioration in the audio quality during digital to digital recording. There are some disadvantages to digital recording such as high end roll-off after 20 kHz (although this is well above the human hearing range), and of course the current high equipment costs.

The digital recording process takes the original analog signal, electronically samples it and changes (encodes) it into a series of electronic pulses by a process called pulse code modulation (PCM). PCM samples the analog signal at rates in the range of 44.1 kHz or 48 kHz, while converting it into a series of discrete voltage levels. Conversely, changing the digi-

tal audio back into the analog format (which must eventually be done in order for us to hear it) requires that the pulses be changed back (decoded) into the original waveform (analog) shape. Thus, each complete digital audio system must contain an analog-to-digital (A/D) converter, the digital audio processing and mixing devices, and a digital-to-analog (D/A) converter.

There are three methods for recording digital audio:

- recording on audiotape
- recording on videotape
- recording to memory disc (commonly called tapeless recording)

Recording on Audiotape

As I previously mentioned, the stationary head configurations and recording formats on today's state-of-the-art digital ATRs tend to vary. For instance, the professional digital (PD) format used by Mitsubishi and Otari is incompatible with the comparable digital audio stationary head (DASH) format used by Sony. In fact, digital audiotape is also different from analog audiotape. For one thing, both the backing and magnetic oxide coating on the digital tape are thinner than the analog tape. This is partly due to the fact that some digital ATRs can run at higher speeds than the

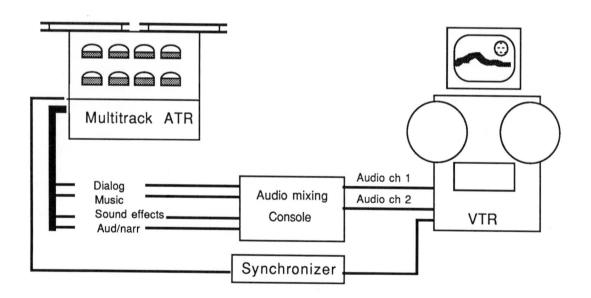

Figure 8.22: Diagram of the audio post-production layback process.

Figure 8.23: The Otari DTR-900 PD format 1-inch digital multitrack audio recorder is available with 24 or 32 tracks. Courtesy Otari Corporation.

Figure 8.24: The Sony PCM-1630 Digital Audio Processor. Courtesy Sony Communications Products Company.

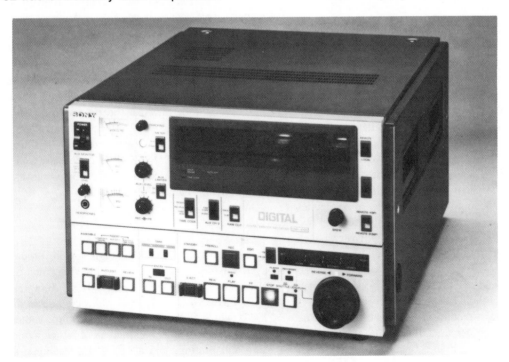

Figure 8.25: The Sony DMR-4000 3/4-inch Digital Master Recorder. Courtesy Sony Communications Products Company.

common 7.5 ips, 15 ips and 30 ips, and therefore require more tape stock per reel. The magnetic oxide coating is also thinner because the constant digital sampling frequency negates the need for the tape to handle the wide range of analog frequencies and amplitudes. The coercivity of digital tape is also much higher than that of analog tape because of the high sampling frequencies involved in the recording. Analog audio frequencies normally range from around 20 Hz to 20 kHz, whereas digital sampling frequencies start at 44.1 kHz and go higher.

Digital audiotape recording is also different from analog audiotape recording because there are several dedicated tracks instead of multiple identical channels. For instance, the Otari DTR-900 PD format, 32 track DATR shown in Figure 8.23, actually records 45 tracks. There are 32 tracks for recording digital audio data, eight parity tracks (two for each set of eight tracks) which serve as error detection and correction tracks in case the digital audio data on one or more tracks is momentarily lost due to dust, dropout or tape damage, two reference analog audio tracks for monitoring and as an aid in cut/splice editing, two auxiliary digital tracks which can store mixing console automation data and one SMPTE/EBU time code track.

Recording on Videotape

The process of recording digital audio on videotape is becoming very popular within the record and music video industry. By using a digital audio processor (DAP), such as Sony Corporation's PCM-1630 shown in Figure 8.24, in conjunction with a 3/4-inch VCR digital master recorder (DMR) like the Sony DMR-4000 shown in Figure 8.25, audio engineers can rerecord the digital music tracks from the multitrack digital ATR through the DAP which converts the signal into a specific code. The coded signal is then recorded by the DMR into the vertical interval of the NTSC video signal. Music video producers and record companies are now starting to record their digital audio and two reference analog audio tracks onto the master 1-inch videotape prior to online editing. This gives them a finished music video with first generation digital stereo audio tracks, and analog stereo audio tracks which can be used if a digital-to-analog converter is not available. The major requirement in this case is that the 1-inch video recorder used for online editing must have the sync record head turned off, otherwise the digital audio tracks will be obliterated during the edit session.

Recording to Memory Disc

The third recording media does not require magnetic tape at all. Instead, the audio is recorded, stored and accessed off large capacity, hard disc drives. This results in the ability to edit and mix audio with true random access. Two such random access systems are the Synclavier® (now in production) and the Opus.

Synclavier®

The Synclavier® digital audio system by New England Digital Corporation (shown in Figure 8.26) will definitely not be a part of every small audio facility in the near future because of its present-day cost of between $100,000 and $300,000 per system. However, some major audio post-production facilities have begun making the investment because of the system's unique features. The Synclavier® can be used for composing music scores, for digitally recording and storing music libraries, for digitally recording and storing dialogue and sound effects, and for printing out music scoring sheets. It can also be used as a keyboard music instrument and as a time-code controlled sound effects machine and editor. Any digitally recorded audio can be stored, accessed, altered and played back in real time. For instance, the Synclavier's 76 keys act as on/off switches; each key controls four different sounds. This makes it possible to have as many as 304 foley and library sound effects active on the keyboard at one time. The system can enter SMPTE/EBU start and stop times in 1/80th of a frame accuracy. Real-time manipulations such as volume adjustments and stereo positioning can be adjusted manually using the keyboard or they can be programmed into the system memory. An entire project's sound effects and music requirements could be accessed, customized, programmed for SMPTE time code start and stop points, and rerecorded in real time onto one or more tracks of the final multitrack ATR.

Opus

The Opus random-access digital audio processing system is designed by Lexicon Inc. This total digital system, which runs in the $100,000 to $200,000 range, will consist of a workstation (like the one in Figure 8.27) which is the audio editor's control center, an electronic cabinet for handling all of the audio processing and routing, and a disc storage cabinet which will contain all of the disc drives.

The Opus will permit 12 channels of digital mixing,

Figure 8.26: The Synclavier® Digital Audio Processing System. Courtesy New England Digital Corporation.

Figure 8.27: The Opus Digital Audio Production System workstation. Courtesy Lexicon Inc.

with up to eight sub-outputs simultaneously recorded or replayed from one of the 800 Mb hard discs. Opus is being designed to edit and process an almost unlimited number of dialogue, music and sound effects tracks, as well as to digitally record ADR and foley effects. As we can see, the workstation will contain aspects of an audio mixing console and an audio edit controller in an effort to make the controls more familiar to most editors and mixers. The audio signals will be processed through the electronic cabinet and will never actually go through the workstation, so it will be possible to have the workstation installed up to 1000 feet away from the cabinets.

The primary audio storage will consist of from one to four 800 Mb hard discs installed in the disc storage cabinet, with each disc capable of holding up to 120 track-minutes of audio. Opus will also feature an optional write-once 1 Gb optical disc subsystem which will be used for source audio backup and archival storage.

CONCLUSION

This ends our discussion of audio post-production for video, and in doing so, we have come to the completion of the video post-production process. Hopefully, reading this chapter has brought about a basic awareness of the audio post-production process. This chapter was not written with the intention of turning each reader into an audio engineer—there are other good books listed in the bibliography for those who want more in-depth coverage of the subject. I merely wish to help the reader understand the process, become more confident and work more efficiently in his or her chosen profession. For this reason I have discussed audio mixing room design and the uses for the various types of audio equipment. In addition, I have tried to stress the organizational aspects of audio post-production, such as spotting cues and material acquisition, because disorganization inevitably leads to substandard audio quality as well as wasted personnel, time and money.

I felt that it was important to include a brief discussion of digital audio recording methods because, while it is still economically out of reach for the majority of small audio post-production facilities, the situation will undoubtedly change (witness the recent proliferation of many lower priced digital video effects devices which we discussed in Chapter 7). I trust from the outset of this book it has been abundantly clear that the video post-production and audio post-production industries are constantly going through rapid change and advancement. No doubt the current processes will change in the next few years as quickly as they have in the past few years.

In our final chapter we shall talk about some of these changes—particularly the emphasis toward digital component video recording, direct film negative to video editing controllers, random-access offline video editing, video editing systems using personal computers and high definition television.

9 The Future: Trends and Directions

I began this book by declaring that we are in the midst of a video revolution. Fueled by rapid, continuing advances in digital circuitry and computer technology, this revolution has transformed the video post-production process. In fact, within a few years, the advanced editing systems and digital special effects devices discussed on the previous pages will probably appear as antiquated as the physical splicing blocks used during the infancy of video post-production.

Although the ultimate direction the video revolution will take is uncertain, one outcome seems clear: the full digitalization of production, post-production and broadcasting facilities. In addition, as the number of commercial, industrial and cable outlets for video programming increases, programs will be produced to appeal to the specialized audiences that these outlets serve. Thus, although the total number of TV productions will be up, the size of the audience for individual programs will be down, resulting in smaller production budgets. Consequently, there will be increased emphasis on producing programming as efficiently and inexpensively as possible, without compromising quality.

ADVANTAGES OF DIGITAL VIDEO

Compared to their older "analog" counterparts, digital components offer faster setup time and greater reliability. In addition, future production crews will be recording entirely "component" rather than "composite" TV signals. In the component approach that was described in Chapter 2, the luminance and color information are recorded separately, thus reducing the problem of signal degeneration that occurs as the video signal proceeds through the post-production and the final delivery stages. In fact, the future may see many TV projects shot, edited, distributed and even transmitted in the component video format, only to be converted at the television receiver, if even then. Sony's Betacam® and Panasonic's MII format VTRs already use the component method of recording, with high-quality results.

Video editing systems of the future will be designed to provide instant access to digitally stored video and audio information, eliminating the time-consuming search and cue functions necessary with today's video editing systems. The old CMX 600 offline editing system achieved this type of random access, but the high price of the system, combined with the necessity of storing relatively small amounts of video and audio information on relatively large disc drives, severely limited its commercial potential. Currently, instant random access VCR and videodisc-based edit systems have become the focus of attention. They show promise in allowing film and video editors the capability to shorten or lengthen edited video sequences without editing down a generation, and, ultimately, to online assemble programs in "actual time" (i.e., assembling an hour-long project using only one hour of online time).

FILM-TO-TAPE TRANSFERS

In the near future, most post-production facilities will also be able to transfer and edit film negatives

Figure 9.1: CMX 6000 random-access offline editing system. Courtesy CMX Corporation.

directly onto videotape, eliminating the need for prior telecine transfers. This will be possible by interconnecting one or more telecine systems with a field-accurate computerized editing controller (such as the time logic controller (TLC) that was discussed in Chapter 4). The gentle film handling, color correction and image manipulation features of the telecine system, combined with the field-rate accuracy of the editing controller, will produce a finished video product only one generation removed from the film negative.

Toward this end, advances in editing technology have begun to allow the interweaving of video and film techniques, both in production and post-production. Many film editors are now beginning to take advantage of the speed and versatility of video editing techniques.

EDIT SYSTEM DESIGN:
THE SEARCH FOR SIMPLICITY

Not only are the future trends in electronic editing design heading toward upscale videodisc-based random access systems, but also toward using the home personal computer at the more affordable end of the scale.

The booming home personal computer industry has already made several advancements possible for the video professional in today's competitive market. New products, such as the Edit Lister™ software program discussed in Chapter 5 and the SC-ED edit control system by Calaway Engineering, use the normal off-the-shelf personal computer to provide video professionals with lower cost features that were available only on high priced, top of the line systems just a few years ago. The popularity of the personal computer "mouse" controller has inspired Videomedia's Mickey—a simple to operate, keyboardless, low priced editing system for cuts only or A/B roll editing.

It is estimated that the majority of television film series produced in the United States are already being electronically edited. Producers' recognition of the speed and economy of electronic post-production for film is providing the incentive for the development of systems such as the CMX 6000 and the Laseredit.

CMX 6000

The CMX 6000 random-access editing system, under development by CMX, is designed to provide the same simplicity of handling pictures and sound as a film flat-bed editing table, but with the advantages of video technology. It will operate as a random-access electronic editing system using laser videodisc, instead of videotape, to access the source material.

The complete system, shown in Figure 9.1, will consist of the edit console, which contains all controls for locating and electronically splicing the source material; the control bridge, which contains the computer, audio monitoring controls and LED program running time display; and the picture and sound modules, which contain two laser videodisc players each, two picture monitors and a keyboard for entering all notations.

The original picture and sound tracks on either film or videotape are first transferred to 30 minute Laservision-compatible videodiscs. The videodiscs are then loaded into the picture and sound modules of the 6000 system to be used as source material. Next, either the editor or the assistant will log the necessary information (scene/take numbers, time code, shot description, etc.) relevant to each scene into the system from the main console or from the stand-alone computer logging station.

As the editing progresses, the 6000 system produces a simulated workprint—nothing is actually recorded. The rapid access time (the discs will scan 30 minutes of material in 1.5 seconds) of each videodisc permits real-time previews of all or selected segments of the edited material. Internal trims and edit point adjustments can be readily accomplished without the need to record down a generation, as is the case with conventional linear video editing. At the completion of editing, any video recorder can be attached to the system to record the simulated workprint playback.

On the basic 6000 system, special effects like dissolves, soft cuts and wipes are notated in the list by using a built-in chinagraph marker using either standard film editing marks and/or by creating two custom marks. If the effects must be seen, an optional effects kit is in the works that will allow the 6000 to interface with certain video effects switchers.

The final edit decision list can be outputted as either a film cutter's log with opticals list for negative conformation, or a CMX list for videotape automatic assembly.

Laseredit

The Laseredit offline video editing system uses laserdisc and 3/4-inch videocassette players to achieve almost instant access of source material while maintaining the linear method of video offline editing.

The typical Laseredit system, shown in Figure 9.2, consists of four Pioneer industrial-grade dual-head

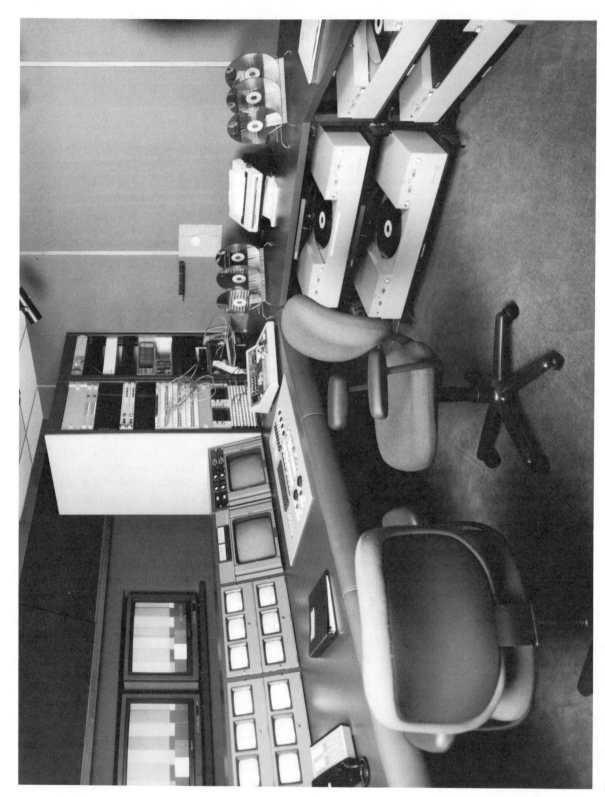

Figure 9.2: Laseredit videodisc-based offline editing system. Courtesy Laseredit, Inc.

laserdisc players, two Sony BVU videocassette recorder/players, a Grass Valley 100 series video production switcher, two color monitors, 12 monochrome monitors and a Spectra Image audio mixing panel. The system is capable of handling up to 11 single-head laserdisc players or five dual-head laserdisc players since it treats each head as a separate source VTR.

The secret of Laseredit's success lies in the high speed, dual-head laserdisc players which are capable of scanning the entire 30 minute disc in two thirds of a second. The client's original source material (usually 1-inch videotape) is reproduced onto the relatively inexpensive, 30 minute plastic sandwiched discs at Laseredit's in-house affiliate Spectra Image, using disc recorder units manufactured by Optical Disc Corporation (ODC), which also manufactures the disc blanks. Each ODC disc is mastered as a CAV (constant angular velocity) type so the disc rotation speed remains constant no matter where the pickup heads are located on the disc. This is particularly crucial when using the dual-head player units since each disc is accessed and treated as two sources. In other words, the editor can cut or dissolve to different locations on the same disc as if it were two separate sources.

Each disc contains video, two channels of audio, SMPTE time code, disc frame identification and the sum data of time code information recorded at the end of the disc for confidence checking. When each disc is loaded onto the player, the system fast scans it to memorize its time code information including source recording stop/starts and code breaks, then it reads the sum data at the disc end to confirm its findings. Consequently, the original source master recording does not require continuous time code or even sequential time code. In fact, the time code could also be a mixture of drop frame and non-drop frame code if it is unavoidable.

The system works extremely well for single camera or multiple camera film or video projects where the offline workprint copies must be made after production shooting. The process's cost efficiency factor would be somewhat reduced on any productions that already utilize the simultaneous recording of workprint cassettes as part of their production deal.

Laseredit's quick accessing time permits the editor to view any take almost instantly, thereby speeding up the decision making process—the major time element in offline editing. The selected takes are edited onto a 3/4-inch videocassette in the normal, linear editing manner so that the result is an edited 3/4-inch videocassette workprint master at the end of the first cut. All subsequent editing requires using the first cut

master videocassette in conjunction with the laserdisc players to create cut two, and so on. The Laseredit software contains its own list cleaning and tracing programs to compile a final edit decision list for online auto-assembly. The online auto assembly at the Laseredit facility can be done in the conventional A mode audio/video editing process, or as in the case of many film-to-video projects, a video-only edit decision list can be auto-assembled on 1-inch while the time-coded Nagra 1/4-inch production audio tapes are auto-assembled onto a 24 track audio recorder by an audio-only edit decision list. This process is favored by many filmmakers because it allows them to have more film-style control over the audio post-production process.

As of early 1988, Laseredit systems are available for use on an hourly basis at Laseredit's Burbank, CA facility. They can also be leased off premise on a short term (7 to 13 weeks minimum) or a long term (more than one year) basis. In the future, Laseredit's designers plan to convert their product into a totally disc-based, random access system as soon as it becomes feasible to do so without the costly necessity of either making multiple copies of each disc or using bank upon bank of laserdisc players.

SC-ED Editing System

The considerable appeal of Calaway Engineering's SC-ED and SC-ED Plus keyboard editing control system, shown in Figure 9.3, is that it operates with the user's existing IBM-PC or Compaq Desk Pro home computer with at least 256K of memory and four free slots to be used for the communications, edit support and general purpose interface (GPI) modules. The installation of the SC-ED or SC-ED Plus editing control package does not affect the normal operation of the computer in business applications, so there are no equipment warranty problems to worry about. In addition, the installation, which consists of direct plug-in control system and GPI modules, can be done by the user in less than 15 minutes.

IBM-PC owners may operate up to any combination of four RS-422 direct control Sony VTRs with internal time code readers, or any combination of four parallel control VTRs, by installing the optional upST Sony Translators which make the SC-ED think that each VTR has RS-422 serial control protocol. This provides the same function as the interface described in Chapter 3. In addition, the system will control various types of audio and video mixing devices, plus it contains several GPI switches for triggering any equipment not capable of being direct or interface controlled.

Figure 9.3: SC-ED Plus video editing controller. Courtesy Calaway Engineering.

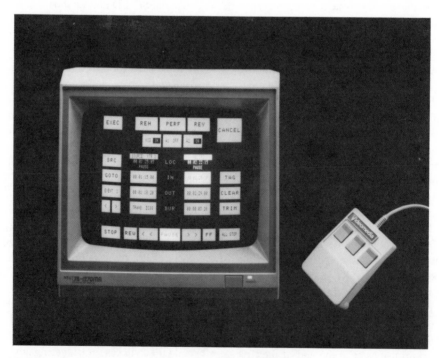

Figure 9.4: Mickey video editing controller. Courtesy Videomedia.

The SC-ED's initialization menu provides the user with a comprehensive choice of editing conditions that can either be preset or customized to the user's specific equipment and skill level. SC-ED also offers a motion controller package for providing variable speed shuttle and frame jogging for those VTRs that are capable of the function. Recently released SC-ED versions of the 409 and trace list management programs, plus an optional 8-inch disc drive unit (replacing the 5 1/4-inch unit) allow the user to interchange EDLs done on either the CMX or GVG editing systems.

Editors using a Compaq Desk Pro can upgrade to the SC-ED Plus package which features considerably faster list management and edit control functions, in addition to operating up to any combination of six VTRs.

Many editors refer to Jack Calaway's edit control package as CMX or GVG transparent since its operation is very similar and it does virtually all of the functions of the high level systems—at a much lower cost.

Mickey

Another move toward blending personal computer-type technology with user friendly simplification, is Videomedia's cuts-only model Mickey I and its A/B roll upgrade model Mickey II. As Figure 9.4 indicates, this is a unique editing controller because there is no control panel or keyboard unit. All of the operating instructions are input by a two-button mouse control interface that highlights and activates the appropriate boxes on the monitor display menus. It is very simple to begin operating the system—the editor simply moves the mouse over the first display, called the Executive Form and highlights each appropriate box. Clicking the left-hand button on the mouse activates the function. This is the initialization portion of the system start-up procedure which determines such items as the individual VTR readiness status, the pre-roll and post roll times, assembly or insert edit modes and list printing options. The final step is to highlight the edit box and click the mouse button which changes the executive form, display to the edit form display on the monitor. The Edit Form display contains all of the boxes needed for the edit session such as VTR selection, VTR motion control, event in and out marks, video and/or audio edit selection and so forth.

The Mickey I and II systems contain a memory capacity of 50 events that can either be printed out after each event or when the list is completed. VTR control interfaces called transport control processors (TCPs) must be installed on each VTR used in the system. The Mickey I cuts-only system will control one source VTR and one record VTR. The Mickey II A/B roll system will control two source VTRs and one record VTR. In addition, the system controls a built-in video and two channel audio dissolve unit capable of nine different selectable frame durations.

In the near future, a PC link will become available for talking to IBM-PCs using the Edit Lister™ program that we discussed earlier in Chapter 5.

ELECTRONIC IMAGERY: HIGH DEFINITION TELEVISION

High definition television (HDTV) is a revolutionary new concept introduced and demonstrated by Nippon Hoso Kyokai (NHK), Japan's national broadcasting corporation. Working in cooperation with Panasonic (a subsidiary of Matsushita Electric Industrial Co., Ltd.), Sony Corp. and Ikegami, Tsushinki Co., Ltd., NHK designed an HDTV system that employs special cameras, videotape recorders, monitors and a large screen projection system. The NHK system is based on a scan rate of 1125 lines, compared to the 525 lines scanned in NTSC systems. HDTV offers a wide screen aspect ratio of 5 x 3, compared to the 4 x 3 aspect ratio of U.S. television.

In January 1982, CBS, Inc. predicted that high definition, wide screen television with stereo sound would establish a presence in home television broadcasting by the late 1980s. CBS also predicted the rapid growth of ''electronic cinema'': high quality feature ''films'' produced on HDTV systems. Still in the research stage is a laser-based system for transferring high definition TV imagery to 70mm film.

Far from causing the demise of the motion picture industry, these innovations should make it much more economical and efficient to produce and distribute features on either video or film. High definition electronic imagery will cut costs by shortening the lead time necessary for feature production and by greatly decreasing the length of post-production.

In the future, many movie theaters may be equipped to receive high definition video features transmitted from direct broadcast satellites. Once the necessary reception equipment is in place, satellite distribution will result in considerable cost savings over the old method of mailing or trucking heavy film prints.

DESIGN COMPATIBILITY AND STANDARDIZATION

As the number of components used in video production and post-production increases, the television industry faces a formidable challenge: maintaining compatibility among the various components and systems. Toward this end, the Society of Motion Picture and Television Engineers (SMPTE) has formed a number of standardization study groups—groups comparable to the study committees that successfully developed industry standards for time code and 1-inch VTR systems.

Of particular interest to post-production professionals is a proposal for editing list standards drafted by the SMPTE committee on video recording and reproduction technology. (The proposal is available from SMPTE.) Through this and its other standardization efforts, SMPTE is serving as a vital link between the manufacturers of video equipment and those who actually use the equipment.

CONCLUSION

The emerging trends described in this chapter will affect both video programming and video editing. However, editors will still continue to follow the basic procedures presented in this book for the three main stages of post-production—preparation, offline editing and online editing.

Whatever the future may bring, it will not change the importance of three major themes of this book. First, video editing is a process, a progression that can be broken into a logical sequence of steps. Second, it is both a technical and a creative undertaking. It requires not only a knowledge of editing equipment, but also a feel for the artistic principles that determine how a scene should be pieced together. Finally, and perhaps most important, video editing is a cooperative process—a team activity that requires the involvement and input of the producer, the director, the associate director, the script supervisor, audio technicians, videotape operators and, of course, the video editor.

Appendix A: Broadcast Transmission Standards Worldwide

	VHF Transmission Standard	UHF Transmission Standard
Algeria	PAL-B	—
Argentina	PAL-N	PAL-N
Australia	PAL-B	—
Austria	PAL-B	PAL-G
Bangladesh	PAL-B	—
Belgium	PAL-B	PAL-H
Brazil	PAL-M	PAL-M
Canada	NTSC-M	NTSC-M
Chile	NTSC-M	NTSC-M
China	PAL-D	PAL-K
Denmark	PAL-B	—
Finland	PAL-B	PAL-G
Great Britain	PAL-I	PAL-I
India	PAL-B	—
Indonesia	PAL-B	—
Ireland	PAL-I	PAL-I
Italy	PAL-B	PAL-G
Japan	NTSC-M	NTSC-M
Malaysia	PAL-B	PAL-G
Norway	PAL-B	PAL-G
New Zealand	PAL-B	—
Netherlands	PAL-B	PAL-G
Portugal	PAL-B	PAL-G
Spain	PAL-B	PAL-G
Sweden	PAL-B	PAL-G
Switzerland	PAL-B	PAL-G
Thailand	PAL-B	PAL-G
Turkey	PAL-B	PAL-G
United States	NTSC-M	NTSC-M
Yugoslavia	PAL-B	PAL-G

Appendix B: Selected Video Editing Systems

Manufacturer	Model	Maximum Index Method	Number of VTRS	Equipment Controlled	Prints Hardcopy EDL	Computer Readable EDL Format
Ampex	HPE-1C	SC	4	CDEFG	YES	PD
Ampex	HPE-104	SC	4	CDEFG	YES	D
Ampex	ACE	SC	16	ABCG	YES	PD
Calaway	SC-ED	SC	6	BCDEFG	YES	D
Cezar	Super Controller	C	2	DEF	NO	N
Cezar	Editing Center	SC	3	DEF	YES	P
Cezar	ABR-1	SC	3	DEF	YES	P
CMX	EDGE	SC	3	BCDE	YES	PD
CMX	3400	SC	7	ABCDEG	YES	PD
CMX	3600	SC	7	ABCDEG	YES	PD
Control Video	Light Finger	SC	6	BCDEF	YES	N
Convergence	Super 90	SC	2	DEF	NO	N
Convergence	100 Series	SC	4	BCDEF	YES	PD
Convergence	200 Series	SC	4	BCDEF	YES	D
EECO	EMME	SC	9	BCDEFG	YES	P
EECO	IVES-1	SC	2	DEF	YES	N
Grass Valley Group	Super Edit	SC	7	ABCDEG	YES	PD
Jatex	VSEC-42T	J	2	ABCDEFG	NO	N
Jatex	VSEC-45TDX	CJ	2	CDEF	NO	N
Jatex	VSEC-62TMX-CA	SCJ	5	ABCDEFG	YES	D
JVC	RM-88U	C	2	DEF	NO	N
JVC	VE-90AT	C	2	DEF	NO	N
JVC	VE-92	SC	2	DEF	YES	D
Panasonic	NV-A500	C	3	EF	NO	N
Panasonic	NV-A970	SC	2	EF	NO	N
Panasonic	AU-A70	SC	3	EF	NO	N
Panasonic	AG-A650	C	2	DF	NO	N
Sony	RM-440	C	2	DE	NO	N
Sony	BVE-800	SC	3	CE	YES	P
Sony	BVE-3000	SC	3	CEG	YES	P
Sony	BVE-6000	SC	8	CEG	YES	PD
United Media	Commander 1	SC	8	BCDG	YES	PD
Videomedia	EAGLE	MC	2	D	YES	D
Videomedia	MICKEY	MSC	3	DEF	YES	N
Videomedia	Z6000A-C	MSC	9	BCDE	YES	D
Videomedia	Z6000D-E	MSC	9	ABCDEG	YES	D

Key to Symbols

Index Method	Equipment Controlled	Computer Readable EDL Format
C Control Track	**A 2-inch Quad VTR**	**D Disc**
J Scene-Dex (Jatex)	**B 1-inch B Format VTR**	**N None**
M Micro-Loc (Videomedia)	**C 1-inch C Format VTR**	**P Punch Tape**
S SMPTE time code	**D 3/4-inch U-type VTR**	
	E 1/2-inch Beta VTR	
	F 1/2-inch VHS VTR	
	G Video Switcher	

Appendix C: Selected Manufacturers of Video Editing Equipment

Ampex Corp.
401 Broadway
Redwood City, CA 94063
415-367-2011

Calaway Engineering
49 So. Baldwin Ave.
Sierra Madre, CA 91024
818-355-2094

Cezar. *See* International Video Corp.

CMX Corp.
3303 Scott Blvd.
Santa Clara, CA 95050
408-988-2000

Control Video Corp.
1640 Dell Ave.
Campbell, CA 95008
408-866-7447

EECO, Inc.
1601 E. Chestnut Ave.
Santa Ana, CA 92701
714-835-6000

The Grass Valley Group, Inc.
P.O. Box 1114
Grass Valley, CA 95945
916-273-8421

International Video Corp.
455 W. Maude Ave.
Sunnyvale, CA 94085
408-738-3900

Jatex, Inc.
2626 Freewood
Dallas, TX 75220
214-358-0200

JVC Company of America
41 Slater Drive
Elmwood Park, N.J. 07407
201-794-3900

Paltex
2752 Walnut Ave.
Tustin, CA 92680
714-838-8833

Panasonic Industrial Co.
One Panasonic Way
Secaucus, NJ 07094
201-348-7000

RCA/Commercial Communications Div.
Front and Cooper Sts.
Camden, NJ 08102
609-338-3000

Sony Communications Products Co.
1600 Queen Anne Rd.
Teaneck, NJ 07666
201-833-5200

United Media, Inc.
4075 Leaverton Ct.
Anaheim, CA 92807
714-630-8020

Videomedia, Inc.
211 Weddell Dr.
Sunnyvale, CA 94086
408-745-1700

Glossary

A/B roll: The use of alternating scenes, recorded on separate videotape reels, to perform *dissolves, wipes* or other types of video transitions.

active crossover network: Term for the frequency separating device used in loudspeaker systems when installed prior to the power amplifier. Compare with *passive crossover network.*

address: A precise *frame* location on a videotape identified by *time code* number; also, the location of specific data in a computer memory.

ADO (Ampex Digital Optics): Trade name for the *digital video effects* system manufactured by Ampex Corp.

ADR (automatic dialogue replacement): See *looping.*

AGC (automatic gain control): Electronic circuitry that compensates for either audio or video input level changes.

A mode: Linear method of assembling edited footage; in A mode assembly, the editing system performs edits in a numerical sequence, stopping whenever the *edit decision list* calls for a reel that is not assigned to a VTR. Compare to *B mode.*

Note: Words that appear in italics within definitions are defined elsewhere in this glossary.

analog: A continuously varying electric signal.

analog recording: The common form of magnetic recording where the recorded waveform signal maintains the shape of the original waveform signal.

A-only edit: Audio-only edit.

ASCII: American Standard Code for Information Interchange. The standard that governs the sequence of binary digits in a computerized *time code* or video editing system.

aspect ratio: Numerical ratio of picture height to width; in video, the standard aspect ratio is 3:4.

assemble edit: Electronic edit that replaces all previously recorded material with new audio and video and a new *control track,* starting at the edit-in point.

AST: See *automatic scan tracking.*

ATR: Audiotape recorder.

attenuate: To reduce the power level of an electrical signal.

audio: The sound portion of a program.

audio post-production: See *sweetening process.*

audio sweetening: See *sweetening process.*

aural exciter: An audio device used to enhance the clarity, brilliance, presence and intelligibility of the signal.

auto assembly: Process of assembling an edited videotape on a computerized editing system, under the control of an *edit decision list.*

automatic scan tracking (AST): A special *head* assembly in *helical* VTRs that allows the video heads to move in two planes simultaneously, eliminating tracking errors introduced by faulty *servo systems,* misaligned tape guides or partially stretched videotape.

AUX: See *auxiliary channel.*

auxiliary channel (AUX): In a video editing system, a channel reserved for connecting an external audio and/or video device.

A/V edit: Edit that records new audio and video tracks. Also called both cut.

back porch: Portion of a *composite video* signal between the trailing edge of the *horizontal sync* pulse and the leading edge of the video portion of the signal.

backspacing: Process of rewinding the videotape a precise distance from the desired edit-in point. This allows the VTRs to come up to speed and synchronize before reaching the edit-in point.

back time: Method of calculating the edit-in point by subtracting the duration of the edit from the edit-out point.

balanced line: Three wire processing mode for signals where two signal wires are isolated and the third wire is connected to system ground. Compare with *unbalanced line.*

banding: Visual disturbance characterized by horizontal variations in *video luminance* and/or color. Banding occurs on *quadruplex* videotape machines when the four video channels are improperly adjusted during recording and/or playback.

band pass: Range of frequencies that will pass through a recording or playback device without distortion.

baud: Unit for measuring the rate of digital data transmission. Usually one baud equals one *bit* per second.

bearding: Video distortion that appears as short black lines extending to the right from bright objects in the scene. Bearding is caused by interruptions in the *horizontal sync* of the video signal. Also called overdeviation or tearing.

Betacam®: Broadcast quality 1/2-inch videocassette format developed by Sony Corp.

B format: *SMPTE* designation for a type of *helical* VTR. See also *segmented video.*

biamplification: A system using two power amplifiers to drive split electronic signals, such as a *two-way loudspeaker system.*

binary system: Coding system that uses the values ''0'' and ''1.''

bit: Smallest unit of computer data. See also *baud, byte, word.*

black burst: See *color reference burst.*

blanking: Portion of the *composite video* signal between the active picture segments for making the horizontal and vertical retrace scan lines invisible. See also *horizontal blanking interval, vertical blanking interval.*

B mode: ''Checkerboard'' or nonsequential method of *auto assembly.* In B mode assembly, the computerized editing system performs all edits from the reels that are currently assigned to VTRs, leaving gaps that will be filled by material from subsequent reels.

both cut: See *A/V edit.*

breakup: Disturbance in the picture or sound signal caused by loss of *sync* or by videotape damage.

breezeway: Portion of the video signal's *back porch* between the trailing edge of *horizontal sync* and the start of *color reference burst.*

B roll: Exact copy of the A roll original material, or new original material on a separate reel, for use in *A/B roll* editing.

bumping up: Transferring a program recorded on a narrower videotape to a wider videotape (e.g., from 3/4-inch to 1-inch videotape).

buss: Row of buttons used to select various input signals, such as those used on a *video switcher*.

BVB: Black-Video-Black. A *preview* mode that displays black, newly inserted video, then black again.

B-wind: In video, winding the videotape onto a reel so the *oxide* side of the tape faces out.

byte: Computer term for a group of binary digits operated upon as a single unit. In most editing system computers, one byte equals eight *bits*.

capstan: Rotating shaft on the VTR *tape transport* path. The capstan governs the speed of the videotape as it travels from the supply reel to the take-up reel.

capstan servo system: Electro/mechanical circuitry that allows the edit VTR to synchronize the playback speed of its tape to the incoming video signal. See also *servo system*.

cathode ray tube (CRT): TV picture tube.

central processing unit (CPU): The main processing section of a computer.

C format: *SMPTE* designation for a type of 1-inch *helical VTR*. See also *nonsegmented video*.

C mode: Nonsequential method of assembly in which the *edit decision list* is arranged by source reel number and ascending source time code.

character: Single letter, number or symbol used to represent information in a computer or video program.

character generator: Electronic device that generates letters, numbers or symbols for use in video titles. Also a device that converts electronic *time code* into visible numbers displayed on a TV monitor.

checkerboard assembly: See *B mode*.

chips: *Gray scale* test pattern used for adjusting television cameras.

chroma: Video color.

chroma key: Method of inserting an object from one camera's picture into the scene of another camera's picture by using a solid primary color background behind the object and processing the signals through a special effects generator.

chrominance: *Saturation* and *hue* characteristics of the color television signal; the portion of the TV signal that contains color information.

cinching: Videotape damage due to creasing or folding.

clipping: Imposing an electronic limit to avoid overdriving the audio or video portion of the television signal.

clock time code: See *drop-frame time code*.

coercivity: A measurement of the magnetic force necessary to change a fully saturated magnetic tape to full erasure, expressed in Oersteds.

color background generator: Electronic circuit used in conjunction with a *video switcher* to generate a solid color background of any desired *hue* or *saturation*.

color balance: Adjustment of the primary colors (red, green, blue) to arrive at a desired overall balance.

color bars: Standard color test signal, displayed as rows ("bars") of color.

color black: Video signal containing *horizontal sync, color reference burst* and *pedestal*.

color frame: Condition at edit point where the *color reference burst* phase of the incoming shot is very close to or the same as the phase of the outgoing shot.

colorimetry: Adjustment of primary and secondary colors, without affecting overall balance of the color signal.

color reference burst: Color synchronizing signal inserted on the *back porch* of *horizontal sync*. When compared with the *color subcarrier* signal, the color reference burst determines the *hue* of the video image.

color subcarrier: Carrier frequency (3.58 MHz in *NTSC*, 4.43 MHz in *PAL)* on which the color signal information is impressed.

color time code: See *non-drop-frame time code*.

compander: An audio noise reduction device that compresses the input signal during encoding and expands the output signal during encoding.

component video: Video signal in which the *luminance* and *sync* information are recorded separately from the color information.

composite video: Video signal containing both picture and *sync* information.

compression (audio): Process of reducing the dynamic range of the audio signal.

compression (video): Lack of detail in either the black or the white areas of the video picture due to improper separation of the signal level.

conforming: Transferring *edit decision list* information created during *offline editing* to allow the final assembly of the finished program in *online editing*. See also *auto assembly*.

contrast ratio: Measurement of the darkest to lightest discernible areas of the video signal.

control track: Portion of the video recording used to control the longitudinal motion of the tape during playback.

control track editor: Type of editing system that uses *frame pulses* on the videotape *control track* for reference.

control track pulse: See *frame pulse*.

convergence: Proper merging of points. In video, the point at which red, green and blue signals are properly aligned in the color monitor.

core memory: *Random access memory* device comprised of magnetized ferrite cores.

CPU: See *central processing unit*.

crawl: Graphics information moving either vertically or horizontally through the picture.

crossover frequency: The frequency at which the low and high frequencies are divided and sent to the *woofer* and *tweeter* sections of a loudspeaker system.

cross talk: Interference between two or more adjacent audio channels.

CRT: See *cathode ray tube*.

crystal black: See *color black*.

cue: To instruct the computerized editing system to shuttle a videotape reel to a predetermined location.

cue track: Audio track typically used for *cue* tones, *time code* or some other cueing signal.

dB: See *decibel*.

decibel: Unit of measurement for sound levels.

degauss: To demagnetize (erase) all recorded material on a magnetic video or audiotape.

delay dissolve: Process in which a scene is edited in, plays for a duration, then dissolves to a new scene. See also *dissolve*.

demodulate: To separate a signal from the carrier frequency onto which it was impressed.

differential gain: Change in the *gain* of the *color subcarrier* as the signal increases from black to white amplitude.

differential phase: Change in the *phase* of the *color subcarrier* as the signal increases from black to white amplitude.

digital: Electronic system which functions by converting the *analog* signal into a series of discrete binary bits.

digital audio processor (DAP): A device for converting an *analog* audio signal into a digital code and inserting it into the video signal.

digital audio stationary head (DASH): A *digital recording* format for reel-to-reel digital audio recorders.

digital frame store: Memory device that scans and stores a complete *frame* of video information after the information has been converted from analog to digital form. Digital frame stores are used to create digital special effects.

digital recording: A method of recording in which the *analog* signal is electronically sampled at a high frequency (encoded), converted into a stream of pulses and stored either on magnetic tape or hard disk drive.

disk: See *floppy disk*.

dissolve: A video transition in which the existing image is partially or totally replaced by superimposing another image.

distribution amplifier: Amplifier that allows one video or audio signal to be sent to several pieces of equipment simultaneously.

downstream keyer: Keying amplifier that operates on the output signal of a *video switcher*.

downtime: Time when equipment needed during a project is inoperable.

drop-frame time code: *SMPTE time code* format that skips (drops) two frames per minute except on the tenth minute, so the time code stays coincident with real time. Compare to *non-drop-frame time code*.

dropout: Drop in the playback *radio frequency* level, resulting in a ''dropping out'' or ''flaking'' of portions of the playback video.

dropout compensator: Device used to reinsert picture information lost in video *dropouts*. Also called DOC.

dub: To make a copy of a video recording.

dump: To copy stored computer information onto an external medium such as hard copy, paper tape or floppy disk.

dupe: To duplicate a videotape; same as *dub*. Also a duplicate copy of a tape.

DVE (digital video effects): Trade name for video system sold by NEC.

dynamic range: An audio term which refers to the range between the softest and loudest levels a source can produce without distortion.

EBU: European Broadcast Union.

edge damage: Damage to the edges of a videotape, generally caused by improper tape guide clearance along the *tape transport* path.

edit decision list (EDL): List of edits performed during *offline editing*. The EDL is stored in *hard copy, floppy disk* or *punch tape* form and is used to direct the final *online editing* assembly of the video programs.

edit pulse: See *frame pulse*.

EDL: See *edit decision list*.

EE (electronics to electronics): Electronic signal passing through a device without being affected.

effects re-entry: The ability of larger *video switchers* to select one *mix/effects row* as an input to another mix/effects row.

EFP (electronic field production): Videotape production using single camera film-style shooting technique.

EIA: Electronic Industries Association.

EIAJ: Electronics Industries Association of Japan.

electronic editing: Process of assembling a finished video program in which scenes are joined together without physically splicing the tape. Electronic editing requires at least two VTRs: a playback VTR and a record VTR.

electronic news gathering (ENG): Using portable video equipment to record news events.

electronic scratch pad: Section of a computer editing program used for making calculations.

encoding: Adding technical data such as *time code, cues,* or closed caption information to a video recording.

ENG: See *electronic news gathering*.

enhancing: Electronically adjusting the quality and sharpness of a video image.

equalization: Adjustment of various frequency ranges to achieve a desired sound.

equalizing pulses: *Vertical blanking* pulses used to maintain proper scanning interlace during the *vertical sync* period.

event: Number assigned by the editing system to each performed edit.

fade: Usually, a *dissolve* from full video to black video or from full audio to no audio.

fade lever: Lever, slider or other control that *attenuates* or amplifies the video or *audio* signal. Usually means adjusting from silence (audio) or black (video) to full level, or vice versa.

feedback: Disturbing high pitched sound created by the output of an audio device being fed back to the input. A receding, endless video loop created in the same manner with video devices.

field: Half of one television *frame*. In *NTSC* video, 262.5 horizontal lines at 59.94 Hz; in PAL, 312.5 horizontal lines at 50 Hz.

film chain: A total system, comprised of a film projector, slide projector and video camera, used to convert film or slides to video. Also called telecine.

film-to-tape transfer: The process of recording film images onto videotape, using a *film chain* and VTR combination.

flash: Interference or breakup to one *field* or less of the video signal. Also called a hit.

floppy disk: Flat, flexible magnetic medium used to store data in computer-readable form. In video editing, floppy disks are used to store *edit decision lists*. Compare to *hard copy, punch tape*.

flying erase heads: Moving erase *heads* mounted near or next to the video heads on a VTR. Flying erase heads are activated during an *insert edit*.

flying spot scanner: System for transferring film to videotape in which the electron beam inside a *cathode ray tube* continuously scans the moving film. Flying spot scanner telecine systems are rapidly replacing the mechanical pull-down mechanism systems used in early *film-to-tape transfers*.

foley: The process of recreating sound effects in sync with a video playback.

four-way loudspeaker system: A four-loudspeaker system configuration where the frequencies are split into low bass, high bass, low treble and high treble.

frame: One complete video picture. A frame contains two video *fields,* scanned at the *NTSC* rate of 30 frames per second or the *PAL* rate of 25 frames per second.

frame address: See *address*.

frame cut: Process in which an edit is performed in *sync* with the same video as the previous shot or edit, to extend the shot or to begin a visual transition.

frame lock: Synchronization of the video signal with *SMPTE time code*.

frame pulse: Pulse superimposed on the *control track* signal. Frame pulses are used to identify video track locations containing *vertical sync* pulses.

frame store: See *digital frame store*.

frequency distortion: A condition when the frequencies present at the input of a device are either not present or are not faithfully reproduced at the output of the device.

front porch: Portion of the *composite video* signal that starts at the trailing edge of the picture information and ends at the leading edge of the *horizontal sync*.

full-coat: Magnetic film stock completely covered with a magnetic oxide coating.

gain: Amplitude (strength) of the video or audio signal.

general purpose interface (GPI): An electronic device containing several electronic switches that can be activated by a remote data signal. In computerized editing systems, GPIs allow the computer to control various remote components.

generation: Copy of original video program material. The original videotaped material is the first generation. A copy of the original is a second generation tape, and so on. Generally, the edited master tape is a second generation tape.

generic tape: Usually, an edited master tape that does not include situation-specific titles or tags. This specific information is added later, when a copy of the generic master is prepared for broadcast in individual markets.

geometry error: *Time-base* and *velocity errors* caused by changes to tape, *head* or guide dimensions between the recording and playback of a videotape.

GPI: See *general purpose interface*.

graphic equalization: Audio signal amplification or *attenuation* in fixed frequency ranges.

gray scale: Range of *luminance* levels from black to white. See also *chips*.

guard band: Blank section of tape used to prevent *cross talk* interference between audio record tracks.

hard copy: Printed version of computer *output*. In video editing, the printed, "human-readable" version of the *edit decision list*. Compare to *floppy disk, punch tape*.

hardware: Mechanical, electrical or magnetic equipment used in video recording or editing. Compare to *software*.

harmonics: Frequencies that are exact multiples of the original frequency.

harmonic distortion: A condition where the audio system introduces *harmonics* that were not in the original signal.

head: Magnetic pickup device in a VTR used to record, erase or reproduce video and audio signals.

head contouring: Form of video distortion on *quadruplex* VTRs caused by the rounding of the corners of the four video *heads*.

helical VTR: Type of VTR in which the videotape is wrapped partially or completely around the video head assembly, resulting in the video signal's being recorded across the tape in slanted strips. Also called slant-track VTR.

hertz (Hz): A unit used to measure frequency. One hertz equals one cycle per second.

high-band color: FM carrier frequency (7.06 MHz to 10 MHz) used for modulating and demodulating during recording or playback on *quadruplex* VTRs. High-band color is of higher quality than *low-band color*.

high-energy tape: Videotape with higher *coercivity* and retentivity than normal *oxide* types. Typically, a high-energy tape is coated with cobalt-duped ferric oxide or chromium dioxide.

high pass filter: A device which only passes frequencies above a preset point.

hit: See *flash*.

horizontal blanking interval: Portion of the *composite video* signal between the end of picture information on one horizontal line and the start of picture information on the next horizontal line.

horizontal lock: During the playback of a videotape, the condition that exists when the *horizontal sync* of the playback video is synchronized with the *horizontal sync* of the post-production facility or transmitting station.

horizontal sync: The portion of the *composite video* signal that synchronizes the scanning electron beam of the TV monitor so that each line of picture information will start at the same lateral position during the scanning process.

hue: The shade of a color.

IEEE: Institute of Electrical and Electronic Engineers.

image enhancing: See *enhancing*.

initializing: Process of setting the computer edit program to proper operating conditions at the start of the editing session.

in-point: Starting point of an edit.

input: External information that is entered into a computer.

insert edit: Electronic edit in which the *control track* is not replaced during the editing process. The method of electronic editing in which a new segment is inserted into program material already recorded on the videotape.

interface: Device used to interconnect two pieces of equipment.

interlace scanning: *NTSC* television scanning process in which two *fields* of video are interlaced to create one full *frame* of video.

interlocking: Running and/or recording separate audio and video tracks in synchronization, using either an electronic or mechanical means of signal coordination.

intermodulation distortion: A condition where two or more frequencies simultaneously pass through a device and interact to create unintended *harmonic* tones that were not present in the original signal.

I/O device: *Input/output* equipment used to send information or data signals to and from an editing computer.

IPS: Inches per second.

IRE: Institute of Radio Engineers.

Iso: Isolated camera.

jamsync: Method of *assemble editing* where the *time code* remains continuous across edit points by synchronizing time code generator numbers with the time code on the edited master.

jitter: Jumping or instability in the TV picture, often caused by synchronization or *tracking* errors.

jogging: Process of moving the videotape forward or backward one *frame* at a time.

joystick: Control device that allows editors to shuttle or jog videotape forward or backward, or to keep the tape still-framed. Joysticks also control the 360° manipulation of the video image on digital effects devices.

jump cut: A mismatched edit that creates a visual disturbance when replayed and displayed on a TV monitor.

key: Electronic method of inserting graphics over a scene (luminance key) or of placing one video image into another (*chroma key*).

keystone: Slanting effect in the video image caused by the camera shooting a graphics card from an angle other than straight on.

kinescope: A method for making a film copy of a TV program by pointing a movie camera at a television screen displaying the program. An early method of recording TV programs, made obsolete by videotape recorders.

layback: Transferring the finished audio track back to the master videotape. See also *sweetening process.*

layover: Transferring the edited master audio to the audio room for sweetening.

L-cut: See *split edit.*

LED: See *light emitting diode.*

Light emitting diode (LED): A type of low-level light component used in many frame count and *time code* displays.

limiter: Electronic circuitry used for preventing the audio signal from exceeding a preset limit.

list management: On computer editing systems, a feature that allows the editor to change, trim or shift editing decisions stored in the editing computer's memory. See also *edit decision list.*

load: To transfer data to or from a storage device. See also *input.*

longitudinal time code: Type of *SMPTE time code* that is recorded on the audio track of a videotape. Compare to *vertical interval time code.*

lookahead: Feature of some computer editing systems that allows the editor to precue one VTR to an upcoming edit while another edit is still in progress.

looping: *Audio post-production* process of repeating a scene continuously so that actors can recreate lip-sync dialogue. See *ADR*.

low-band color: FM carrier frequency (5.5 MHz to 6.5 MHz) used for modulating and demodulating during recording or playback on *quadruplex* VTRs. Low-band color is of lower quality than *high-band color*.

low pass filter: A device which only passes frequencies below a preset point.

luminance: Amplitude (strength) of the *gray scale* portion of the television signal.

magnetic disk: See *floppy disk*.

marking: Process of entering *time code* numbers as the videotape is playing in *real time*. Also called "on the fly."

master/slave: Video editing process in which one or more VTRs (slaves) may be directed to imitate the actions of another VTR (master). The process is used in a *real-time edit*.

match frame: See *frame cut*.

matte: Form of *key*. Matte is usually associated with the process of adding electronic color to black and white titles.

megahertz (MHz): One million cycles per second. See also *hertz*.

microphonics: Mechanical vibration of the elements on a video camera's pickup tube, resulting in the spurious modulation and distortion of the normal signal.

microprocessor: A chip or network of chips designed to perform programmed functions in computers and electronic equipment.

microsecond (μsec): One millionth of a second.

millisecond (msec): One thousandth of a second.

mix/effects row: The section of a *video switcher* containing at least two rows of buttons and a *fade lever*, capable of performing *dissolves*, *wipes* and *keys*.

modulation: Impressing a signal onto a carrier frequency for transmission purposes.

moire: In video, a beating pattern produced by harmonic distortion of the FM signal.

monochrome: Black and white.

mortice: Video effect where the picture is compressed and surrounded by a border; normally done in commercial spots.

multiburst: Television test signal used to measure a system's frequency response.

multiple record: Process of recording more than one edited master simultaneously. See also *master/slave*.

multistripe: Film stock containing multiple stripes of magnetic oxide.

NAB: National Association of Broadcasters.

nanosecond (nsec): One billionth of a second.

needle drop: A music library fee rate based on each time part of a selection is used in a program.

noise: Any elements that interfere with the clarity and purity of the TV image.

nonadditive mix: Mixing two images, each of which is set at 100% video level. This contrasts with a *dissolve*, in which the video levels start at 0% for the new image and 100% for the existing image, go through 50%-50% and then end at 100%-0%.

noncomposite video: Video signal that does not contain *horizontal* and *vertical sync* pulses.

non-drop-frame time code: *SMPTE time code* format that continuously counts a full 30 frames per second. As a result, non-drop-frame time code is not coincident with *real time*. Compare with *drop-frame time code*.

nonsegmented video: VTR format in which a full field of video information is recorded and reproduced by one complete scan of the video *head* assembly. Also called unsegmented video. Compare to *segmented video*.

notch filter: A device that passes only a very narrow range of frequencies between upper and lower preset points.

NTSC: National Television Standards Committee, the group that established the color TV transmission system used in the U.S. See also *NTSC color video standard*.

NTSC color video standard: The U.S. standard for color TV transmission, calling for 525 lines of information, scanned at a rate of 30 *frames* per second.

octal system: Numbering system that uses a base of eight. Compare with *binary system*.

offline editing: Preliminary post-production session, used to establish editing points and to prepare an *edit decision list*. Compare to *online editing*.

online editing: Final editing session, the stage of post-production in which the edited master tape is assembled from the original production footage, usually under the direction of an *edit decision list*. See also *auto assembly*.

open-ended edit: Edit that has a start time but no designated stop time.

operating program: Program in an editing computer that allows the videotape editor to control the system's VTRs.

optimizing: Adjusting the VTR record current for proper *radio frequency* amplitude reading.

out of phase: Usually, an edit that is not *color framed*. Also, when the *color reference burst* and *color subcarrier* are not properly adjusted.

out-point: End point of an edit.

output: Information extracted from a computer.

overdeviation: See *bearding*.

overrecord: Edit that is allowed to run longer than planned. All excess frames are eliminated by the next edit.

overscan: Adjustment of a video monitor's picture size so it more closely resembles the picture on a home TV set.

oxide: Metallic coating on videotape that is magnetized during the recording process.

painting: Adjusting the color controls on a video camera.

PAL: Phase Alternating Line, a color TV standard used in many countries. PAL consists of 625 lines scanned at a rate of 25 *frames* per second. Compare to *NTSC color video standard*.

panoramic potentiometer (pan pot): An audio mixing console control that varies the signal level between the left and right channels.

paper edit: Rough *edit decision list* made by screening original material, but without actually performing edits.

paper tape: See *punch tape*.

parametric equalization: Audio signal amplification or *attenuation* in selectable and variable frequency ranges.

passive crossover network: Term for the frequency separating device used in loudspeaker systems when installed after the power amplifier. Compare with *active crossover network*.

patch: Process of connecting two audio or video cables to each other. Also, a temporary correction or adjustment in a computer operating program. See also *patch panel*.

patch panel: Panel on which *input* and *output* connectors are mounted, permitting the temporary interconnection of audio or video components.

peak limiter: Electronic circuitry that prevents sudden peak signal levels from exceeding a preset level.

peak white: Brightest level of the video signal.

pedestal: Reference black level of the video signal (7.5 units for color, 10 units for black and white). Also called setup.

phase: Relative timing of signals. If the signals occur at the same moment, they are in phase. See also *out of phase*.

pinch roller: A rubber roller on the *tape transport* path that presses the videotape against the *capstan*.

pink noise: Generated random noise with a frequency spectrum designed to appear flat across the human auditory hearing range.

prelay: The phase of audio post-production during which music, sound effects, dialogue replacement and announce tracks are added to the master multitrack before the final mix.

preroll: Process of rewinding videotapes to a predetermined *cue* point, so the tapes are stabilized and up to speed when they reach the edit point.

preview: To rehearse an edit without actually performing (recording) it. See also *BVB, VBV, VVV*.

preview row: Row of buttons on a *video switcher* that allows the selected source to be viewed on the preview monitor before committing it to the *program row*.

print through: Transfer or bleeding of a recorded signal from one layer of magnetic tape to adjacent layers, resulting in video or audio distortion.

proc amp: See *processing amplifier*.

processing amplifier: Electronic device that processes the video signals that are fed through it, by allowing adjustment of the signal levels and by providing stable *horizontal* and *vertical sync* pulses.

professional digital (PD): Reel-to-reel digital audio recording format.

program row: Row of buttons on a *video switcher* that sends the selected source out to a *VTR* or transmitter.

protection copy: Duplicate of the edited master reel, kept as a backup in case the master is damaged.

pulse code modulation (PCM): The digital recording process of converting the *analog* signal into a series of digital pulses.

pulse count editor: See *control track editor*.

pulse cross: Display feature on some video monitors that offsets the picture, allowing an editor to see and analyze the *horizontal* and *vertical sync* pulses.

punch tape: Tape punched with holes to represent data in computer-readable form. Used as a method of storing *edit decision lists*. Also called paper punch tape or paper tape. Compare to *floppy disk, hard copy*.

quadrature error: Horizontal displacement of picture elements in the display of a *quad* VTR, caused by a misalignment of one or more of the quad machine's video *heads*.

quadruplex (quad): Videotape system that records with four video heads mounted 90° apart around the rim of a rotating *head* wheel. Quad VTRs generally use 2-inch videotape. Also called transverse scan VTR.

quad split: Special effect in which four different video scenes are displayed simultaneously.

Quantel: Trade name for the digital video effects system manufactured by MCI/Quantel.

radio frequency (RF): High frequency portion of the electromagnetic spectrum used for transmitting television and radio signals. See also *UHF, VHF*.

RAM: See *random access memory*.

random access memory (RAM): Computer memory system that allows users to store and retrieve information rapidly.

raster: Area of the TV picture tube that is scanned by the electron beam. Also, the visual display present on a TV picture screen.

reader/punch: Peripheral device that both reads and generates (punches) *edit decision lists* on *punch tape*.

real time: Actual clock time in which events occur.

real-time edit: Function available on some computerized editing systems that allows the synchronizing of multiple VTRs and the performing of multiple edits in *real time*. See also *master/slave*.

recall: Retrieving a previously performed edit decision from the computer's memory, allowing for corrections or reedits.

reframe: Action taken by a *time base corrector* to correct disturbances in the recorded signal of a *helical VTR*.

registration: Process of aligning the electron guns in a color video camera or color TV to ensure that the electron beam scans the proper points on the *target* (in a video camera) or *raster* (in a TV monitor).

replay: Playing back a previously performed edit.

resolution: Amount and degree of detail in the video image, measured along both the horizontal and vertical axes.

reverb: Reverberation; electronic audio effect resembling an echo.

RF: See *radio frequency*.

RGB: Red, green and blue, the primary colors in color video.

ripple: Process in *edit decision list* management of adjusting the record times of all edits following a length altered edit.

router: Electronic switching system for audio and video signals.

safe action area: The area that will safely reproduce on most TV screens; 90% of the screen, measured from the center.

safe title area: The area that will produce legible titles on most TV screens; 80% of the screen, measured from the center.

saturation: Amount of color in the television picture.

scallop: Irregularity in the TV picture that appears as a repeating curvature at the edge of the image. Scalloping is usually caused by a misalignment of the tape guides on a *quadruplex* VTR.

scanner assembly: Mechanical drum assembly that contains the rotating *head* tips and tape guiding elements used to record and play back *helical* videotape recordings.

SECAM: Sequential Colour a Memoire, a TV standard developed by the French and used in France, the USSR and other countries.

segmented video: VTR format that requires two or more passes of the video *head* to record a full field of video information. Compare to *nonsegmented video*.

selective synchronization (sel sync): A multitrack audio recorder feature which permits changing the record head into a playback head so previously recorded material can be monitored and new material recorded in sync.

serial time code: See *longitudinal time code*.

servo system: Motor system in VTRs responsible for maintaining proper tape speed.

set in/out: Process of entering eight-digit time code by manually typing the digits on the computerized editing system's keyboard.

setup: See *pedestal*.

shading: Process of adjusting the video camera for optimum picture quality.

shedding: Extreme separation of *oxide* from the videotape backing.

signal-to-noise ratio: Measure of the amount of unwanted *noise* present in an audio or video signal. Usually, the ratio of noise introduced by a video or audio component when the output signal is compared with the input signal.

skew: Irregularity in the TV picture caused by improper tape tension or by a difference in *head* tip penetration during recording and playback on *quadruplex* VTRs. Skew usually appears as a saw-tooth effect at the edge of the image.

slant track VTR: See *helical VTR*.

slewing: Process of synchronizing the VTRs in computerized editing systems.

slide chain: See *film chain*.

smearing: Overloading of picture levels, resulting in images flaring or trailing off toward the side of the frame. Smearing is caused by malfunctioning or overloading of the TV camera's clamping circuitry.

SMPTE: Society of Motion Picture and Television Engineers.

SMPTE time code: Binary time code denoting hours, minutes, seconds and frames. See also *time code, binary system*.

software: Computer and video programs and their associated documentation. Compare to *hardware*.

soft wipe: *Video switcher* wipe effect that has diffused edges.

solo switch: An audio mixing console feature that allows individual or groups of channels to be fed to monitoring headphones.

spatial: Term referring to the relative positioning of sounds in a stereo mix.

split edit: Edit in which the audio and video signals are given separate *in-* and/or *out-points,* so the edit takes place with one signal preceding the other. This does not affect the audio and video synchronization.

split screen: Video special effect displaying two images that are separated by a horizontal or vertical wipe line.

spotting: The process of viewing a *window dupe* of the video edit master in order to accurately locate the start and stop points for music, sound effects, *ADR* and narration.

Squeezoom: Trademark for the digital video effects system marketed by Vital Industries.

standing waves: A condition where certain reflected frequencies continue vibrating against parallel surfaces instead of decaying naturally.

still store: Electronic device using *digital frame stores* for storing a series of video freeze frames.

streaking: Video disturbance that appears as dark streaks extending toward the right side of the picture. Streaking is usually caused by faulty clamping circuitry in a video processing device.

subcarrier: See *color subcarrier*.

submaster: High-quality copy of a master tape used to make additional copies. See also *dupe, dub*.

super: The superimposition of one video signal on another, achieved by manipulating the fader controls of a *video switcher*. See also *dissolve, nonadditive mix*.

sweetening process: Process of mixing sound effects, music and announcer audio tracks with the edited master tape's audio track. Also called *audio post-production* for video.

sync: Abbreviation of synchronization. Usually refers to the synchronization pulses necessary to coordinate the operation of several interconnected video components. When the components are properly synchronized, they are said to be "in sync."

sync generator: Component that generates sync pulses.

sync roll: Synchronizing and rolling VTRs for editing purposes. See also *real-time edit*.

sync word: Portion of *SMPTE time code* that indicates the end of each frame and the direction of tape travel. See also *time code*.

tail slate: Slate information recorded at the end of the take; usually recorded upside down.

tape transport: Wheels, gears and motor mechanism that guide and move the magnetic tape through an audio or videotape recorder. See also *capstan, pinch roller, servo system*.

target: Photosensitive front surface of a video camera's pickup tube.

TBC: See *time base corrector*.

telecine: See *film chain, film-to-tape transfer*.

test tone: Continuous or series of reference audio frequencies recorded at OVU at the head of the videotape reel at the start of each recording session.

three-stripe: Magnetic film stock containing three rows of magnetic oxide coating.

three-way loudspeaker system: A three-loudspeaker system configuration where the frequencies are split so that there is one bass range plus low and high treble ranges.

tied masters: Two or more short program segments (usually commercials) recorded consecutively on a length of videotape.

time base corrector: Electronic device used to correct video signal instability during the playback of videotape material.

time base error: Variation in the stable relation of picture information, color information and video sync pulse during the VTR playback process. See also *sync*.

time code: Electronic indexing method used for editing and timing video programs. Time code denotes hours, minutes, seconds and frames elapsed on a videotape. See also *binary system, longitudinal time code, SMPTE time code, vertical interval time code*.

Trace: Trade name for the edit decision list management program marketed by Grass Valley Group, Inc.

tracking: Speed and angle at which the tape passes the video *heads*.

tracking edit: Zero duration edit used as a reference during transition (*dissolve, wipe,* etc.) edits on computerized editing systems.

transverse scan VTR: A *quadruplex* VTR.

trim: Process of adjusting edit points after *time code* numbers have been entered into the editing system.

tweeter: A slang term for the speakers that reproduce the high (treble) frequencies.

two-pop: Short burst of audio tone on a film leader, signifying two seconds before the start of the program audio.

two-way loudspeaker system: A two-loudspeaker system that splits the frequencies into one bass range and one treble range.

UHF: Ultra high frequency TV signals. The signals received as channels above number 13 on a TV set. Compare to *VHF*.

U-matic: See *U-type VTR*.

unbalanced line: Two wire processing mode where one wire carries the signal and one wire is connected to system ground. Compare with *balanced line*.

underscan: Display feature on some video monitors that allows viewers to see the four complete edges of the video signal. Compare to *pulse cross*.

unsegmented video: See *nonsegmented video*.

usable picture area: See *safe title area*.

user bits: Portions of the *SMPTE time code* left blank for adding user information.

U-type VTR: Videocassette recorder format that uses 3/4-inch videotape. See also *helical VTR*.

vacuum guide: Part of the *video head* assembly on *quadruplex* VTRs that is used to maintain the videotape in the correct position relative to the head wheel.

VBV: Video-Black-Video. A *preview* mode that shows a previously recorded scene, a black segment, then the previously recorded scene again.

vectorscope: Electronic test equipment that displays the color information in the video signal.

velocity compensator: Device used to correct *velocity errors*.

velocity error: Rate of change in the *time base error*.

vertical blanking interval: Period during which the TV picture goes blank as the electron beam returns (retraces) from scanning one *field* of video to begin scanning the next. The vertical blanking interval is sometimes used for inserting *time code*, automatic color tuning and captioning information into the video signal.

vertical interval time code: *Time code* that is inserted in the *vertical blanking interval*. Compare to *longitudinal time code*.

vertical sync: *Sync* pulses that control the vertical field-by-field scanning of the *target* area by the electron beam.

VHF: Very high frequency TV signals. The signals received as channels 2-13 on a standard TV set.

video gain: *Gain* of the video signal.

video head: Electromechanical device in a VTR that is used to record and play back video information. See also *head*.

video switcher: Electronic equipment used to switch among various video inputs to a record VTR.

videotape: Oxide-coated, plastic-based magnetic tape used for recording video and audio signals.

VITC: See *vertical interval time code*.

voltage controlled amplifier (VCA): An amplifier whose gain is controlled by a DC voltage.

VTR: Videotape recorder.

VU meter: Volume unit meter. An instrument used to measure audio levels.

VVV: Video-video-video. A *preview* mode that shows a previously recorded scene, new insert video and then the previously recorded scene again.

waveform monitor: A type of test equipment used to display and analyze video signal information.

whip: Horizontal picture disturbance at an edit point, usually caused by timing misadjustments in the electronic editor.

white clip: Circuit that corrects positive overmodulation of a *composite video* signal.

white level frequency: Frequency of the FM signal corresponding to the peak white level of the video signal.

white noise: Generation of random frequencies with even or flat frequency response across the audio spectrum.

window dupe: Copy of an original master recording that features character-generated *time code* numbers inserted in the picture. Window dupes are used in *offline editing*.

wipe: Special effect *transition* in which a margin or border moves across the screen, wiping out the image of one scene and replacing it with another.

woofer: A slang term for the speakers that reproduce the low (bass) frequencies.

word: Computer term for a unit of data comprised of 8 or 16 bits; most current editing computers use 8-bit words.

workprint: Edited master recording created during *offline editing*.

write: See *dump*.

zero duration dissolve: Method of editing two scenes end-to-end simultaneously.

Bibliography

BOOKS

Alten, Stanley R. *Audio in Media* - 2nd edition, Belmont, CA: Wadsworth Publishing Co., 1986.

Anderson, Gary H., *Electronic Post-Production: The Film to Video Guide*. White Plains, NY: Knowledge Industry Publications, Inc. 1986.

Bartee, Thomas C. *Digital Computer Fundamentals*. New York: McGraw-Hill Book Co., 1972.

Bensinger, Charles. *The Video Guide*. Santa Fe, NM: Video Information Publications, 1979.

Efrein, Joel. *Videotape Production and Communication Techniques*. Blue Ridge Summit, PA: TAB Books, Inc., 1970.

Ennis, Harold E. *Television Broadcasting: Equipment, Systems, Operating Fundamentals*. Indianapolis, IN: Howard W. Sams and Co., Inc., Publishers, 1979.

Hubatka, Milton C. and Fredrick Hull and Richard W. Sanders. *Audio Sweetening for Film and TV*. Blue Ridge Summit, PA: TAB Books Inc., 1985.

Kybett, Harry. *Video Tape Recorders*. Indianapolis, IN: Howard W. Sams and Co., Inc. Publishers, 1978.

Mascelli, Joseph V. *The Five Cs of Cinematography*. Cine Graphic Publications, 1965.

Reisz, Karel and Gavin Millar. *The Technique of Film Editing*. New York: Hastings House, Publishers, Inc., 1968.

Robinson, J.F. and Stephen Lowe. *Videotape Recording*. Woburn, MA: Focal Press, 1981.

Tremaine, Howard M. *Audio Cyclopedia*. Indianapolis, IN: Howard W. Sams and Co., Inc., Publishers, 1978.

Utz, Peter. *Video Users Handbook*. White Plains, NY: Knowledge Industry Publications, Inc., 1982.

TECHNICAL MANUALS AND JOURNAL ARTICLES

AVR-3 Videotape Recorder Manual. Ampex Corp., 1980.

BVH 1100 Videotape Recorder Manual. Sony Corp., 1979.

CMX-340X Operator Manual, Revision A. CMX/Orrox, 1981.

"Digital Television." *Broadcast Management/Engineering* 17:6 (February 1981).

DPE 5000 Plus Manual. MCI/Quantel, 1981.

"Electronic Cinema in the Wings." *Variety 194:34 (January 25, 1982).*

"The First VTR: A Historical Perspective." *Broadcast Engineering* 23:5 (May 1981).

"A 4 Billion Dollar Business." POST, (December 1986).

1441 Virsignal Deleter/Inserter Manual. Tektronix, Inc., 1973.

1420 Vectorscope Manual. Tektronix Corp., 1975.

Lund, Robert. "SMPTE Committee on Video Recording and Reproduction Technology: Working Group on Editing Procedures." *SMPTE Journal* 91:9 (September 1982).

MK II DVE Manual. Grass Valley Group, Inc., 1980.

PC II High Speed Reader/Punch Manual. Digital Equipment Corp., 1974.

Remex Papertape Reader/Perforator System Manual. Ex-Cell-O Corp., 1979.

Rosenthal, Eric and Robert Plath. "Television Studios of the Future." *SMPTE Journal* 91:1 (January 1982).

25 Years of Videotape. 3M Corp., 1981.

VPR-2B Videotape Recorder Manual. Ampex Corp., 1981.

Index

ABOUT THE AUTHOR

Gary H. Anderson began his videotape editing career at KCOP-TV, an independent station in Los Angeles, and later worked for Trans-American Video and Vidtronics, Inc. He is currently a videotape editor at Unitel Video, Inc., Hollywood, CA. A member of the Academy of Television Arts and Sciences, he serves on the Academy Blue Ribbon panel for judging single and multiple camera editing awards. He is also a member of the Society of Motion Picture and Television Engineers and the International Alliance of Television and Stage Employees. Anderson is the author of *Electronic Post-Production: The Film-to-Video*

Guide, published by Knowledge Industry Publications, Inc.

Anderson's editing credits include network series such as, ''Don Kirchner's Rock Concert,'' ''Burt Sugarman's Midnight Special,'' ''That's My Mama,'' ''Soap,'' ''Benson'' and ''Family Ties,'' as well as numerous pilots, documentaries, musical productions, commercials and music videos. He has won four national Emmy awards and six nominations for outstanding videotape editing, plus a Monitor award nomination for Best Editor—Broadcast Entertainment.